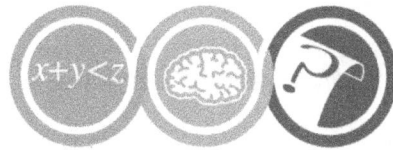

Center for Algebraic Thinking

Teaching Algebra
with
iPad Apps

Doug Neill, Steve Rhine,
and the Center for Algebraic Thinking team

Published 2015 by The Center for Algebraic Thinking

Teacher Algebra with iPad Apps
Copyright © 2014 by The Center for Algebraic Thinking

Address inquiries to:
The Center for Algebraic Thinking
2766 Weatherford Ct NW
Salem, Oregon 97304
algebrathinking@gmail.com

Neill, Doug
Rhine, Steve
Teacher Algebra with iPad Apps

Printed in the United States of America.
First Printing: October 2014
Second Printing: June 2015

ISBN-13: 978-0-692-46824-1
ISBN-10: 0692468242

Table of Contents

Introduction

Algebra is the "gatekeeper" to higher education and future employment because, rather than helping students develop mathematical competence and gain access to higher education, it screens out many students.(1) For instance, in the Los Angeles School District, "48,000 ninth-graders took beginning Algebra; 44% flunked, nearly twice the failure rate as in English... Among those who repeated the class in the spring, nearly three-quarters flunked again...It triggers dropouts more than any single subject".(2) "In Grand Rapids Public Schools, nearly 40 percent of algebra 1 students – 1,196 of 3,149 – received a failing or "held' grade. And 22 percent of the 2,101 algebra II students also did not pass."(3) Responding to Chicago's increasing Algebra failure rate, a school official said "It's not surprising that you're going to see an increase in [failure] rates [in Algebra] if you raise the instructional requirements and you don't raise supports".(4)

Many school districts are responding to the algebra crisis by creating two or three-year algebra programs or having students take double periods of algebra, but the student success rate is not changing significantly.(5) It is becoming apparent that the amount of time spent studying algebra is not the issue. We cannot prepare future teachers to teach algebra the same way and expect different results. Institutions currently working on the algebra issue include the Connected Mathematics Program, College Preparatory Mathematics Program, and West Ed. There is some evidence that new curricula from projects like these are improving students' success with algebra,(6) but many students still struggle to understand.

The National Mathematics Advisory Panel (2008) recommends "teachers should understand how students learn to solve equations and word problems and causes of common errors and conceptual misunderstandings."(7) Research has documented students' mental hurdles in making the transition from arithmetic to algebra.(8) Over 800 articles spanning the last three decades examine why students struggle in algebra. The apps created by the Center for Algebraic Thinking are largely based on that research.

The activities that follow were created to help teachers effectively use the apps developed by the Center in a classroom setting. These simple apps typically have students manipulate and explore a dynamic between variables. Virtual manipulatives can be powerful tools for learning, providing significant gains in achievement.(9) However, "one reason that educational software has not realized its full potential to facilitate and encourage students' mathematical thinking and learning is that it has not been adequately linked with research."(10) Accordingly, the Center is using research on algebraic thinking as the basis for manipulative development. For example, one of our manipulatives addresses research by Monk (1992) that students tend to draw graphs that imitate reality, such as a hill, regardless of the labels of the axes ('iconic translation'). The Action Grapher app shows a bike climbing various hills while simultaneously three separate graphs of height, distance, and speed versus time appear alongside. The student draws what she thinks each graph will look like, then animates the bike and compares her hypotheses against the actual graphs that unfold. These are apps that take little time to use in class and focus on a specific, challenging concept or misconception. Using manipulative computer simulations has resulted in significant learning gains.(11) We believe that the most value comes when those apps are used within the context of a lesson rather than simply handing the iPad to a student and saying, "Here, play around with this.". Through this resource book, we hope to provide you with useful activities and resources to accomplish that goal.

The activities found in this resource fall within three broad categories: pre-app activities, during-app activities, and post-app activities. First, the pre-app activities. These activities are designed to introduce a concept or a game format to students before using the app so that when they do engage with the app, they are prepared to make the most use of their time with the app. These activities are a way to set the stage for what is to come so that they understand how the use of the app fits in with the bigger-picture goals of the lesson, unit, or course.

The during-app activities fill two roles. One is to provide an analog tool to complement the digital nature of the app. That could take the form of simple record-keeping or a separate workspace to do pencil and paper thinking (both of which could serve as accountability checks as well). The second role is to fill in the gaps when there are not enough iPads in the classroom for all students to be using one. Those activities give students not currently using an iPad the ability to engage in a similar activity addressing the same content.

The post-app activities provide opportunities to dig deeper into the content after the app has been explored. They are a way for students to take what they learned while using the app and apply that knowledge in a new way to reinforce what they learned.

In conjunction with each activity we have included a teaching guide to give direction to the instructor regarding how to incorporate the activity into a lesson and why it might be a good idea to do so. Each app is identified with the applicable <u>Common Core</u> <u>Standards</u>.

As we mentioned above, each app is based on research on students' algebraic thinking. A group of 17 math teachers and educators read those 800 articles about students' thinking in algebra. They distilled that research into information that is usable by teachers into <u>The Encyclopedia of Algebraic Thinking</u>. Each entry in the Encyclopedia discusses mathematical issues underlying students' struggles and ways of thinking about algebra, statistics about how often you might expect certain types of answers and misconceptions, and transcripts of how students might sound as they talk about their thinking. In this resource book, each app is linked to entries within the Encyclopedia. We encourage you to explore that Encyclopedia and take advantage of the ability to add comments from your own teaching experiences related to any specific entry.

As we continue developing these resources and creating more apps, we would appreciate your input both on how you are using these in your classrooms, and what others apps or types of apps you would like to see us develop. For feedback and suggestions, please send an email to <u>algebrathinking@gmail.com</u> or connect with us on Twitter (@AlgebraThinking).

Thank you for using our apps to help your students open the gate to higher education. We hope this contribution makes that process a bit more feasible.

<u>Notes</u>

1. Ladson-Billings, 1998; Moses & Cobb, 2001
2. Helfand, 2006, p. 1

3. Murray, 2010
4. Viadero, 2009, ¶8
5. Nomi & Allensworth, 2009
6. i.e. Moseley & Brenner, 2009; Riordan & Noyce, 2001
7. National Mathematics Advisory Panel, 2008, p. 4-88
8. Booth, 1984; MacGregor & Stacey, 1997; and Moseley & Brenner, 2009
9. Crawford & Brown, 2003; Reimer & Moyer, 2005; Suh & Moyer-Peckenham, 2007
10. Sarama & Clements, 2008
11. Roschelle, et al., 2007; Heid & Blume, 2008

References

Booth, L. (1984). *Algebra: Children's Strategies and Errors.* Windsor, UK: NFER-Nelson.

Crawford, C., & Brown, E. (2003). Integrating Internet-based mathematical manipulatives within a learning environment. *Journal of Computers in Mathematics and Science Teaching, 22*(2), 169-180.

Heid, M. & Blume, G. (2008). Technology and the development of algebraic understanding. In M. Heid & G. Blume (Eds.). *Research on technology and the teaching and learning of mathematics.* Information Age Publishing, 55-108.

Helfand, D. (2006, January 30). A formula for failure in L.A. schools. Los Angeles Times. Retrieved from http://www.latimes.com/news/education/la-me-dropout30jan30,1,2605555.story.

Ladson-Billings, G. (1998). Teaching in Dangerous Times: Culturally Relevant Approaches to Teacher Assessment. *Journal of Negro Education, 67*(3), 255-267.

Monk, S. (1992). Students' understanding of a function given by a physical model. In G. H. E. Dubinsky (Ed.), *The Concept of Function: Aspects of Epistemology and Pedagogy.* USA: Mathematical Association of America.

Moseley, B. & Brenner, M. E. (2009). A Comparison of Curricular Effects on the Integration of Arithmetic and Algebraic Schemata in Pre-Algebra Students. *Instructional Science: An International Journal of the Learning Sciences, 37*(1), 1-20.

Moses, R. & Cobb, C. (2001). Organizing Algebra: The Need to Voice A Demand. *Social Policy, 31*(4), 4-12.

Murray, D. (December 20, 2010). Algebra in elementary school? As demands change, educators look for new ways to teach math. *The Grand Rapids Press.* Retrieved from: http://www.mlive.com/news/grand-rapids/index.ssf/2010/12/algebra_in_elementary_school_a.html.

National Mathematics Advisory Panel (2008). *Foundations for Success: The Final Report of the National Mathematics Advisory Panel.* Retrieved from: http://www2.ed.gov/about/bdscomm/list/mathpanel/report/final-report.pdf.

Nomi, T. & Allensworth, E. (2009). "Double-Dose" Algebra as an Alternative Strategy to Remediation: Effects on Students' Academic Outcomes. *Journal of Research on Educational Effectiveness, 2*(2), 111-148.

Reimer, K., & Moyer, P. S. (2005). Third-graders learn about fractions using virtual manipulatives: A classroom study. Journal of Computers in Mathematics and Science Teaching, 24(1), 5-25.

Riordan, J.E., & Noyce, P.E. (2001). The impact of two standards-based mathematics curricula on student achievement in Massachusetts. Journal for Research in Mathematics Education, 32(4), 368-398.

Roschelle, J., Tatar, D., Shechtman, N., Hegedus, S., Hopkins, B., Knudsen, J., Stroter, A. (2007). *Can a technology enhanced curriculum improve student learning of important mathematics?* Technical Report: SRI International.

Sarama, J., & Clements, D. H. (2008). Linking research and software development. In G. W. Blume & M. K. Heid (Eds.), *Research on technology and the teaching and learning of mathematics: Volume 2, cases and perspectives* (pp. 113-130). New York: Information Age Publishing, Inc.

Suh, J. M., & Moyer-Packenham, P. S. (2007). Developing students' representational fluency using virtual and physical algebra balances. *Journal of Computers in Mathematics and Science Teaching. 26*(2), 155-173.

Viadero, M. (2009, March 11). Algebra-for-All policy found to raise rates of failure in Chicago. *Education Week.* Retrieved from http://www.edweek.org/ew/articles/2009/03/11/24algebra.h28.html?tmp=1587018900.

Index Of Common Core State Standards

Grade 6

The Number System

- Two-Player Card Clutter
- Card Clutter Progress Monitor
- Student-Created Card Clutter
- Hop the Number Line: An Introduction
- Hop the Number Line: Student Hop Challenge
- Hop the Number Line: Three Approaches
- A New Football Field
- Inequality Kickoff Progress Monitor

Expressions and Equations

- Algebra Equation Builder Card Game
- Algebra Equation Builder Progress Monitor
- Sticky Note Cover Up Math
- Cover Up Math App Progress Monitor
- Sticky Note Cover Up Challenge
- Comparison of Methods: Cover Up VS Algorithmic
- The Mental Math Cover Up Challenge
- A New Football Field
- Inequality Kickoff Progress Monitor
- Drag and Drop Linear Equations
- Multiple Representation of Lines
- Line Building Challenge
- Moving from Points to Lines
- Moving from Points to Lines: The Follow Up

Grade 7

The Number System

- Algebra Equation Builder Card Game
- Algebra Equation Builder Progress Monitor
- Hop the Number Line: An Introduction
- Hop the Number Line: Student Hop Challenge
- Hop the Number Line: Three Approaches

Expressions and Equations

- Algebra Equation Builder Card Game
- Algebra Equation Builder Progress Monitor
- Sticky Note Cover Up Math
- Cover Up Math App Progress Monitor
- Sticky Note Cover Up Challenge
- Comparison of Methods: Cover Up VS Algorithmic
- The Mental Math Cover Up Challenge

- A New Football Field
- Inequality Kickoff Progress Monitor
- Submariner Algebra Workspace
- Submariner Algebra: The Paper Version

Grade 8

The Number System

- Two-Player Card Clutter
- Card Clutter Progress Monitor
- Student-Created Card Clutter

Expressions and Equations

- Filling the Flask
- The Flask Challenge
- The Doodle Challenge
- The Doodle Duel
- Doodle Pictionary
- Sticky Note Cover Up Math
- Cover Up Math App Progress Monitor
- Sticky Note Cover Up Challenge
- Comparison of Methods: Cover Up VS Algorithmic
- The Mental Math Cover Up Challenge
- Discovering the Mystery Function
- Submariner Algebra Workspace
- Submariner Algebra: The Paper Version

Functions

- Discovering the Mystery Function
- Drag and Drop Linear Equations
- Multiple Representation of Lines
- Line Building Challenge
- Moving from Points to Lines
- Moving from Points to Lines: The Follow Up
- Function Investigation
- Graphing the Terms of Polynomials
- Submariner Algebra Workspace
- Submariner Algebra: The Paper Version
- Tortoise and the Hare Algebra Challenge Progress Monitor

High School: Algebra

Seeing Structure in Expressions

- Algebra Equation Builder Card Game
- Algebra Equation Builder Progress Monitor
- Sticky Note Cover Up Math
- Cover Up Math App Progress Monitor
- Sticky Note Cover Up Challenge
- Comparison of Methods: Cover Up VS Algorithmic

- The Mental Math Cover Up Challenge
- Deriving The Diamond Method - Part 1
- Deriving The Diamond Method - Part 2

Creating Equations

- Algebra Equation Builder Card Game
- Algebra Equation Builder Progress Monitor
- Discovering the Mystery Function
- A New Football Field
- Inequality Kickoff Progress Monitor
- Drag and Drop Linear Equations
- Multiple Representation of Lines
- Line Building Challenge
- Moving from Points to Lines
- Moving from Points to Lines: The Follow Up
- Function Investigation
- Submariner Algebra Workspace
- Submariner Algebra: The Paper Version
- Tortoise and the Hare Algebra Challenge Progress Monitor

Reasoning with Equations & Inequalities

- Sticky Note Cover Up Math
- Cover Up Math App Progress Monitor
- Sticky Note Cover Up Challenge
- Comparison of Methods: Cover Up VS Algorithmic
- The Mental Math Cover Up Challenge
- Discovering the Mystery Function
- Drag and Drop Linear Equations
- Multiple Representation of Lines
- Line Building Challenge
- Moving from Points to Lines
- Moving from Points to Lines: The Follow Up
- Function Investigation
- Submariner Algebra Workspace
- Submariner Algebra: The Paper Version
- Tortoise and the Hare Algebra Challenge Progress Monitor

High School: Functions

Interpreting Functions

- Filling the Flask
- The Flask Challenge
- The Doodle Challenge
- The Doodle Duel
- Doodle Pictionary
- Deriving The Diamond Method - Part 1
- Deriving The Diamond Method - Part 2
- Drag and Drop Linear Equations
- Multiple Representation of Lines
- Line Building Challenge
- Moving from Points to Lines

- Moving from Points to Lines: The Follow Up
- Function Investigation
- Graphing the Terms of Polynomials
- Submariner Algebra Workspace
- Submariner Algebra: The Paper Version

Building Functions

- Discovering the Mystery Function
- Moving from Points to Lines
- Moving from Points to Lines: The Follow Up
- Function Investigation
- Graphing the Terms of Polynomials
- Submariner Algebra Workspace
- Submariner Algebra: The Paper Version
- Tortoise and the Hare Algebra Challenge Progress Monitor

Linear, Quadratic, & Exponential Models

- Funding Your Dreams
- How Often To Compound?
- The Compound Interest Estimation Game
- Discovering the Mystery Function
- Drag and Drop Linear Equations
- Multiple Representation of Lines
- Line Building Challenge
- Moving from Points to Lines
- Moving from Points to Lines: The Follow Up
- Function Investigation
- Graphing the Terms of Polynomials
- Submariner Algebra Workspace
- Submariner Algebra: The Paper Version
- Tortoise and the Hare Algebra Challenge Progress Monitor

Trigonometric Functions

- Function Investigation

Center for Algebraic Thinking

TEACHING GUIDE

Assignment Type: Screenshot Presentation

Overview:

Students demonstrate proficiency by taking screenshots of accomplishments and compiling those screenshots into a presentation to share with the instructor and/or the class.

Common Core State Standards:

The Common Core State Standards that will apply to this activity are specific to the app with which this activity is used.

Encyclopedia of Algebraic Thinking:

The relevant entries of the Encyclopedia of Algebraic Thinking are specific to the objectives of each individual assignment.

Description:

Rather than having students turn in a worksheet as a record of their work, this type of assignment incorporates the tools available on the iPad to allow students to demonstrate knowledge in a more interactive way - they compile their own evidence of accomplishment and present those accomplishments in presentation form. The outline of this process is as follows:

- The instructor provides the objectives. Students must be able to meet these objectives using an app on the iPad.
- For each objective that is met, the student takes a screenshot to demonstration completion.
- The student compiles those screenshots into a presentation (using the Keynote app, for example), adding text and other media to the presentation as needed.
- The student shows/emails the presentation to the instructor or presents it to the class.

Extensions:

In addition to demonstrating completion of objectives, students could add additional slides to discuss difficulties faced while meeting those objectives and how they overcame those difficulties.

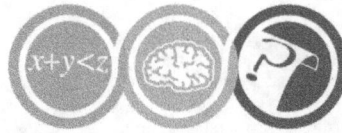

Center for Algebraic Thinking

Developing Game Strategies

For use with any of the apps developed by the Center for Algebraic Thinking

Name: Date: Period:

Name of app: _____

1. Summarize the challenge that you are faced with while playing this game (include a sketch of the game environment):

2. What is *your* strategy for succeeding at this task?

3. Talk with classmates about the strategies they use while playing this game. How are their strategies different than yours? Whose strategies are more effective, yours or theirs? Why?

4. Play the game again, and see if you do any better now that you have more strategies to use!

Center for Algebraic Thinking

Teaching Guide

Developing Game Strategies

App: (Any)

Overview:

Students describe the app, outline their own strategies for succeeding at the task at hand, and share their strategies with their classmates.

Common Core State Standards:

The Common Core State Standards that will apply to this activity are specific to the app with which this activity is used.

Encyclopedia of Algebraic Thinking:

The relevant entries of the Encyclopedia of Algebraic Thinking are specific to the app with which this activity is used.

Description:

The purpose of this activity is to encourage students to pay particular attention to the way in which they approach a task. This activity also incorporates the social aspect of learning - the students experiment on their own and then share strategies with each other - which develops longer-lasting skills than they would if the teacher explained "the one best method" prior to the students beginning the activity on their own.

Extensions:

A follow-up assignment could be for students to monitor their success both before they discuss their strategies with classmates and after, and then reflect on the process.

Action Grapher

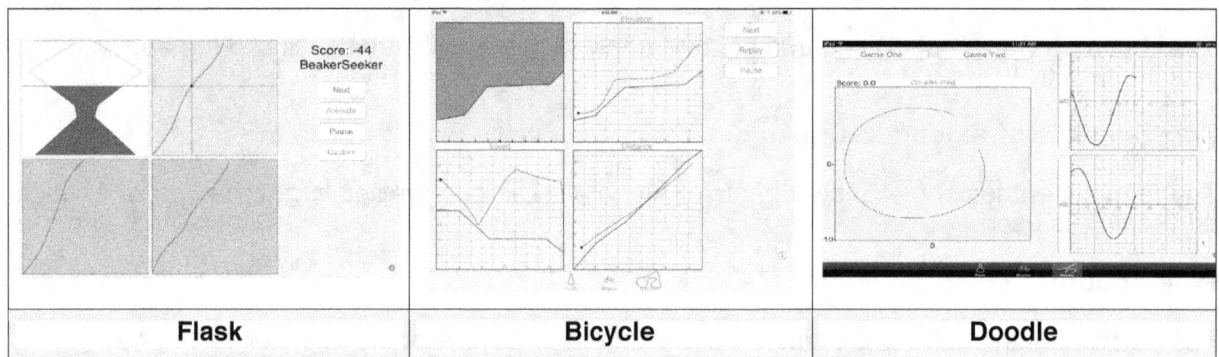

| Flask | Bicycle | Doodle |

Three apps in one! Students often have difficulty understanding the relationship between axes in a graph and how the two variables interact. They tend to believe the graph will look like reality. To develop a more comprehensive understanding of the dynamic taking place in graphs, each of these apps challenge the student to explore how the information from each axis influences the graph. In Bicycle, the user takes a bike on a journey up and down a hill and sees different graphical representations of what is going on based on considering different variables. In Flask, the user draws a flask of any shape, watches it fill up with water, and sees how the graph is influenced by the shape of the flask.In Doodle Pad the user can draw an action and instantly watch two distinct graphs appear, demonstrating different representations of what is happening mathematically.

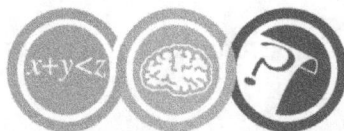

Center for Algebraic Thinking

Filling the Flask

For use with the iPad app *Action Grapher*

Name: Date: Period:

Part 1: Explore

Navigate to the "Flask" tab of the iPad app *Action Grapher*. Read the instructions, and then see if you can select the height vs. time graph that corresponds to the given flask. Also try creating your own custom flasks. Once you can consistently select the correct graph for many different flasks, you can move on to Part 2 below.

Part 2: Explain

Now that you have had a chance to explore this section of the app, your next task is to explain the relationship between the side view of the flask and the graph of height vs. time. To begin, sketch out both the flask and the height vs. time graph of a non-custom flask (i.e., a flask that is automatically generated by the app):

Flask   **Height vs. Time Graph**

Now add a horizontal dotted line to your sketch of the flask to separate unique sections of the flask. Then add a *vertical* dotted line to each unique section of the height vs. time graph. Finally, number each section of your flask sketch and height vs. time graph such that Section 1 of your flask sketch corresponds to Section 1 of your height vs. time graph.

Use the table below to explain what is going on in each section that you labeled on the previous page:

Section	What is happening with the flask in this section?	What is happening with the height vs. time graph in this section?	Do the changes in the flask and in the graph make sense together? Explain.

Teaching Guide

Filling the Flask

App: Action Grapher

Overview:

Students develop and solidify the concept of rate of change as it pertains to the change in the height of liquid in an automatically-generated (not student-drawn) flask as a function of time (at a constant rate of pouring).

Common Core State Standards:

- CCSS.Math.Content.8.EE.B.5
- CCSS.Math.Content.HSF.IF.B.4
- CCSS.Math.Content.HSF.IF.B.6

Encyclopedia of Algebraic Thinking:

- Analysis of Change: Interpreting Graphs
- Analysis of Change: Graphs as a Literal Picture

Description:

This activity will serve as a good introduction to the Flask section of the app Action Grapher. After students are given time to freely explore this section of the app in Part 1 of the activity, in Part 2 they must explain in detail the relationship between the two different representations - the side view of the flask and the graph of height vs. time. You might want to go through one example of Part 2 as a class. After seeing an example students should be able to follow the prompts on their own - with particular attention directed to the unique sections of the height vs time graph - so that they can form a complete and consistent view of the connection between the two representations.

Extensions:

- The Flask Challenge

Center for Algebraic Thinking

The Flask Challenge

For use with the iPad app *Action Grapher*

Name: Date: Period:

This is a Screenshot Presentation assignment. Here are your objectives:

Objective 1:
Navigate to the "Flask" tab of the iPad app *Action Grapher*. For each of the following, draw a custom flask that will produce a height vs. time graph that is the same as the graph shown:

a)

b)

c)

d)

e)

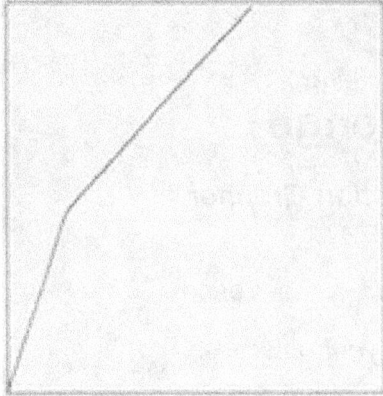

f)

g)

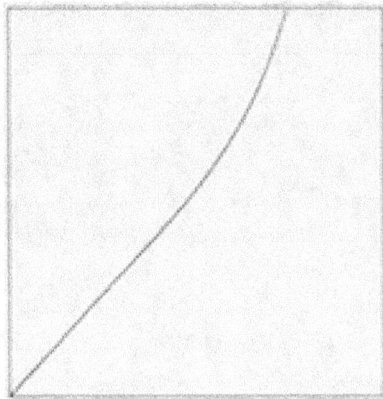

h)

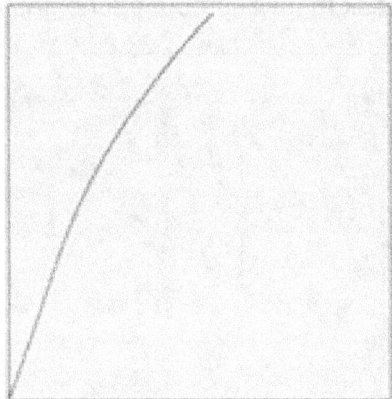

i)

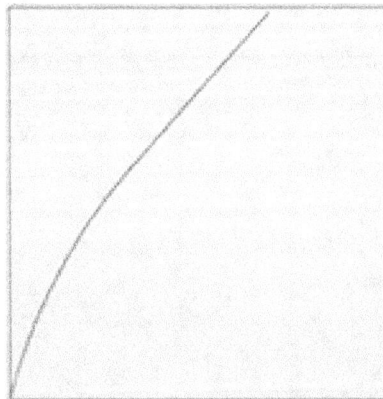

j)

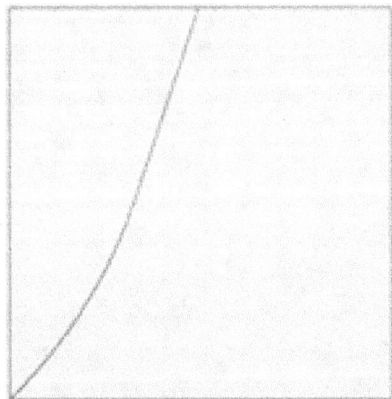

k)

l)

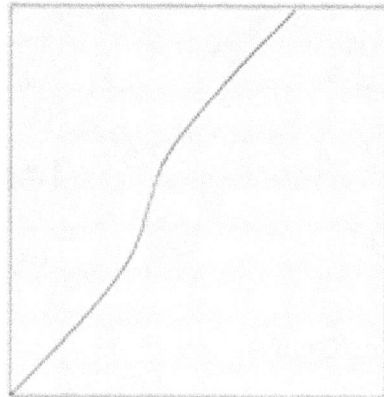

m)

n)

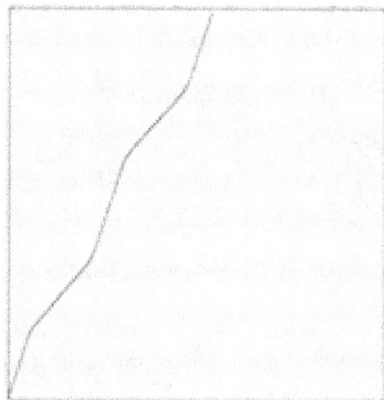

o)

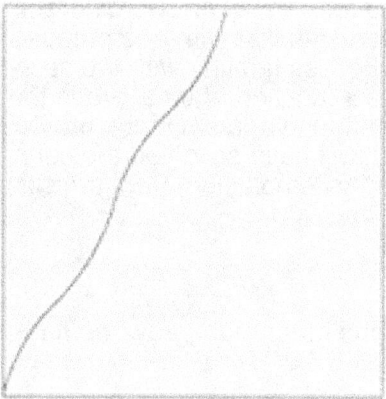

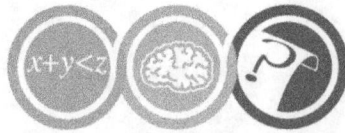

Center for Algebraic Thinking

Teaching Guide

The Flask Challenge

App: Action Grapher

Overview:

Students draw liquid containers in the particular shape that will produce the given height vs. time graph when the container is filled at a constant rate of pouring.

Common Core State Standards:

- CCSS.Math.Content.8.EE.B.5
- CCSS.Math.Content.HSF.IF.B.4
- CCSS.Math.Content.HSF.IF.B.6

Encyclopedia of Algebraic Thinking:

- Analysis of Change: Interpreting Graphs
- Analysis of Change: Graphs as a Literal Picture

Description:

In this activity students will get practice switching back and forth between two different representations of rate of change. Students are likely already familiar with one of those representations - the filling of a liquid container. They will be less familiar with the second representation - a graph of liquid height in the container as a function of time (when the container is filled at a constant rate of pouring). There is enough variation in the provided graphs of height vs. time that after completing all of them students will have solidified the connection between the two representations. The graphs are ordered by complexity - from the simplest to the most complex.

Extensions:

Two-Player Challenge: one students draws h vs. t graph, the other must draw the flask that produces that graph.

Center for Algebraic Thinking

The Doodle Challenge

For use with the iPad app *Action Grapher*

Name: Date: Period:

This is a Screenshot Presentation assignment. Here are your objectives:

Objective 1:
Navigate to "Doodle" tab of the iPad app *Action Grapher*. From the Home page (not Game 1 or Game 2), draw a doodle that produces the following x vs. arc length and y vs. arc length graphs:

a)

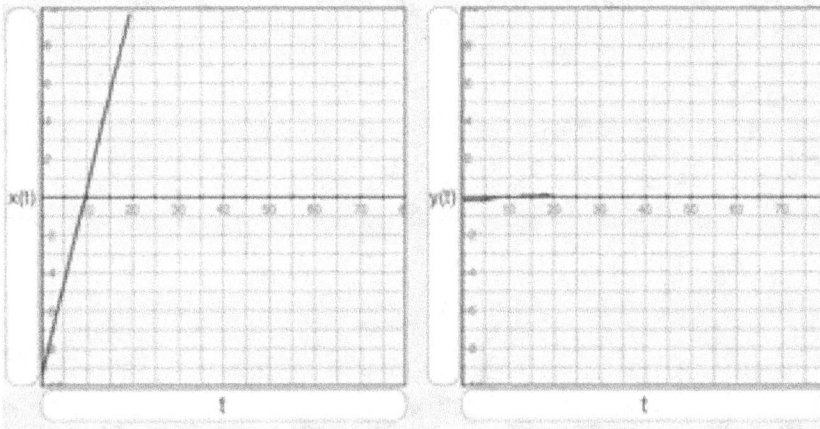

b)

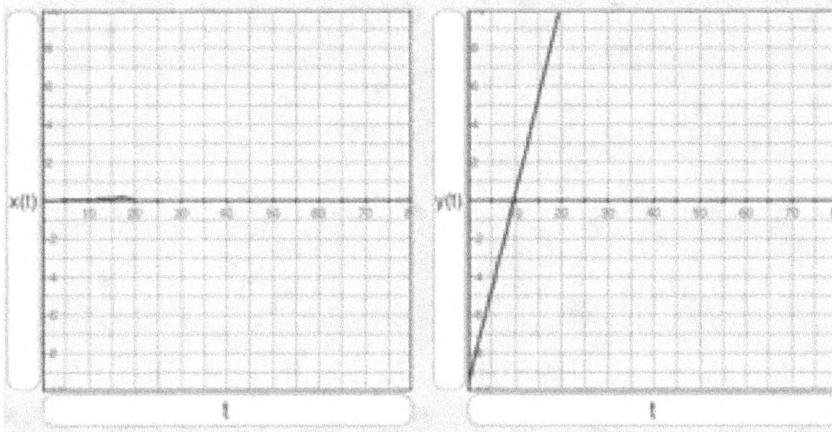

c)

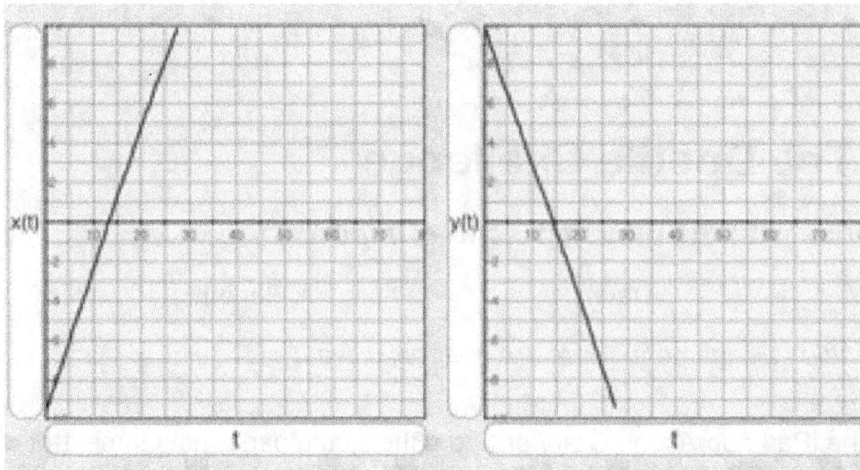

d)

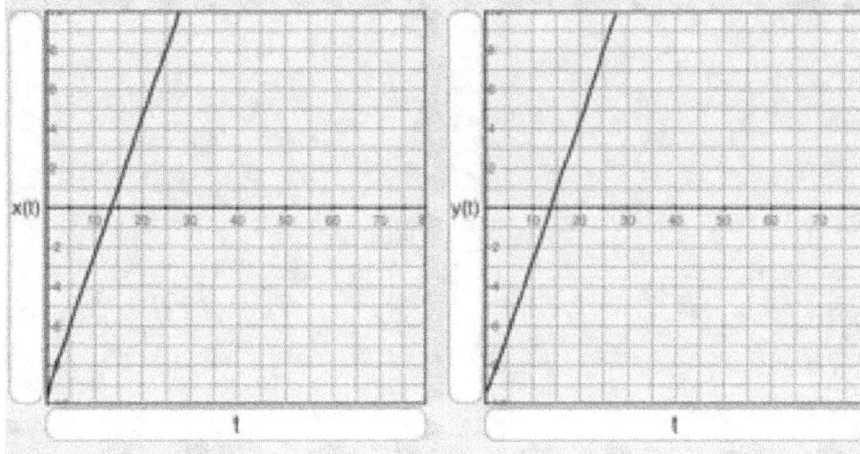

e)

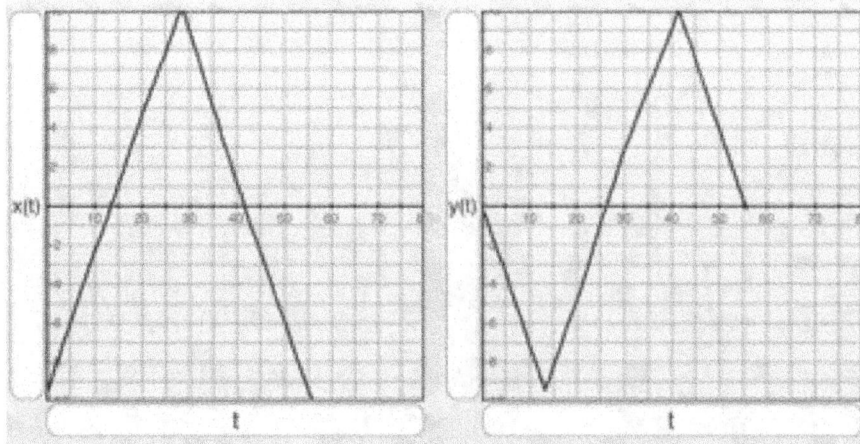

f)

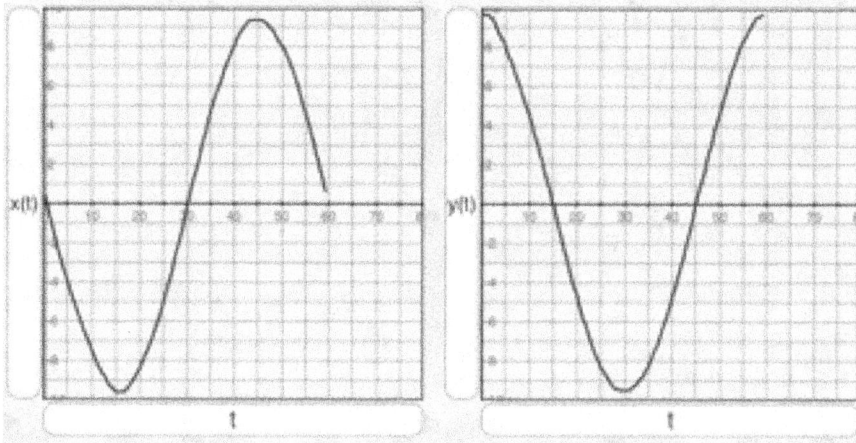

g)

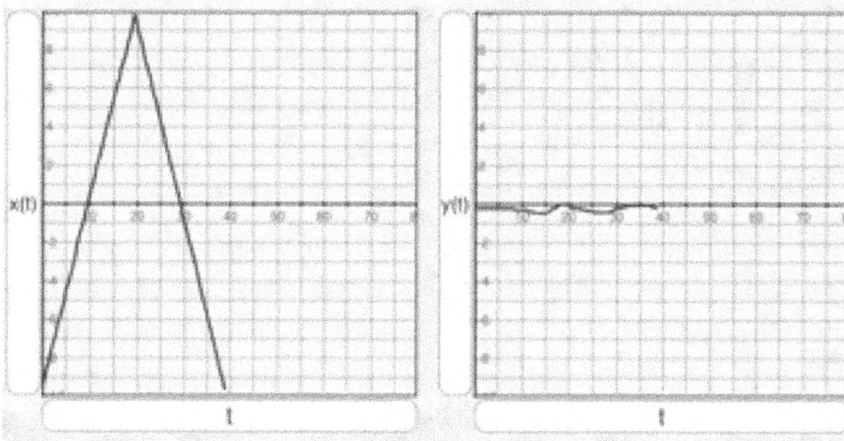

h)

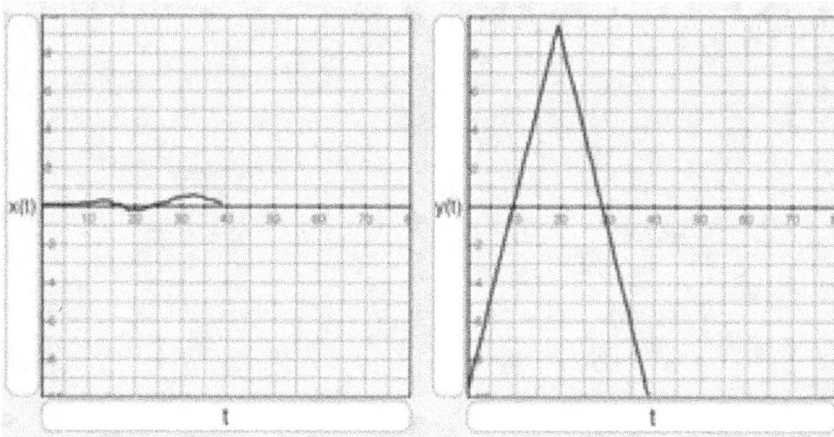

i)

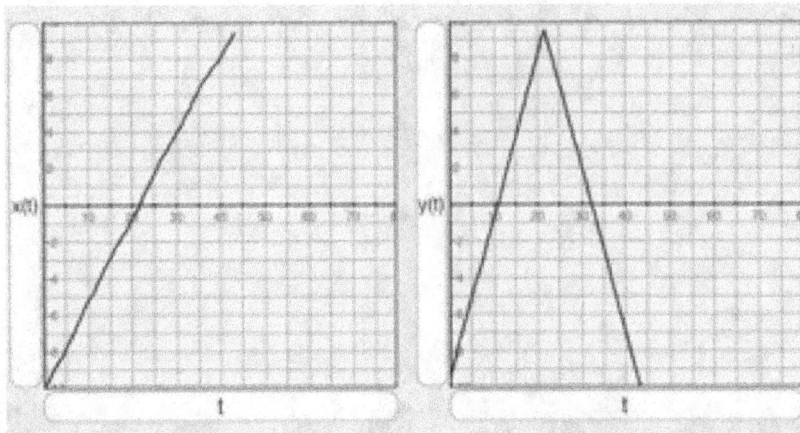

j)

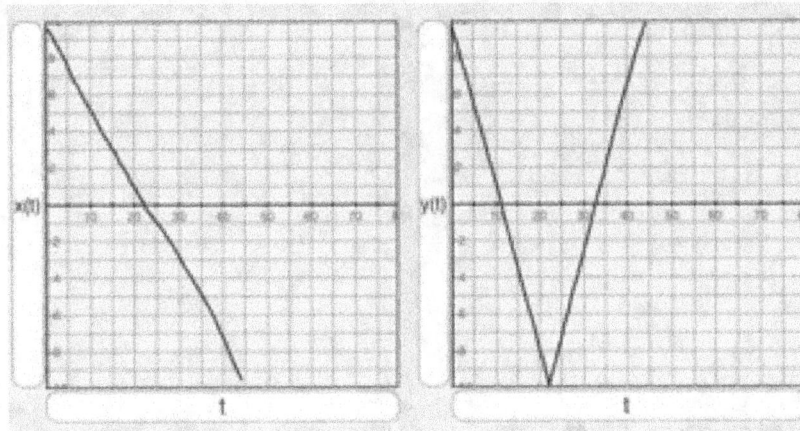

k)

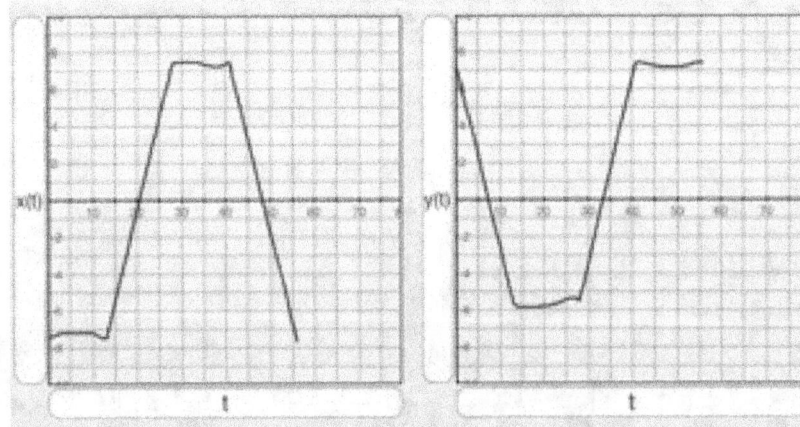

Center for Algebraic Thinking

Teaching Guide

The Doodle Challenge

App: Action Grapher

Overview:

In this Screenshot Presentation assignment, students test their understanding of the Doodle section of the app Action Grapher by attempting to create a doodle that produces the given x vs. arc length and y vs. arc length graphs.

Common Core State Standards:

- CCSS.Math.Content.8.EE.B.5
- CCSS.Math.Content.HSF.IF.B.4
- CCSS.Math.Content.HSF.IF.B.6

Encyclopedia of Algebraic Thinking:

- Analysis of Change: Interpreting Graphs
- Analysis of Change: Graphs as a Literal Picture

Description:

The uniqueness of the Doodle section of the Action Grapher app is that it produces x and y graphs that are a function not of time but of arc length. This will force students to take a conceptual leap into a representation that they likely will not be familiar with. By challenging students to draw two-dimensional doodles that produce the given graphs, they will develop an understanding of the mapping of one representation to the other.

Extensions:

- The Doodle Duel
- Doodle Pictionary
- Game Two of the Doodle section of Action Grapher

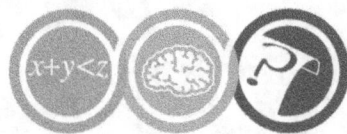

Center for Algebraic Thinking

The Doodle Duel

For use with the iPad app *Action Grapher*

Name: Date: Period:

With a partner, try out this two-person challenge:

- Step 1: Player one draws a doodle that is no longer than the agreed upon length limit and then sketches the x vs arc length and y vs. arc length graph in the space provided below.
- Step 2: Player one clears the doodle from the screen, then hands the iPad and the sketched graphs to player two.
- Step 3: Player two attempts to draw a doodle so that the graphs match those produced by player one. If player two succeeds within the agreed upon time limit, then player two gets a point. If not, player one gets a point. You can keep track of points using tallies in the space provided below.
- Step 4: Players switch roles and then repeat Steps 1 through 3

Length Limit: _____   Time Limit: _____

Player One Score: _____   Player Two Score: _____

Player One Doodle:

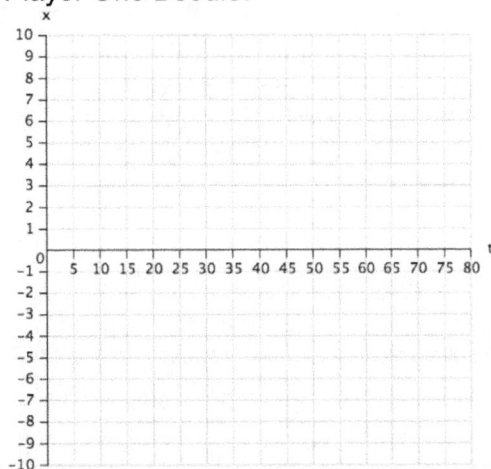

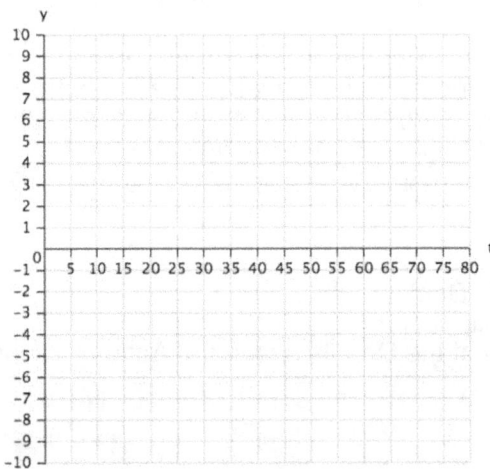

Player Two Doodle:

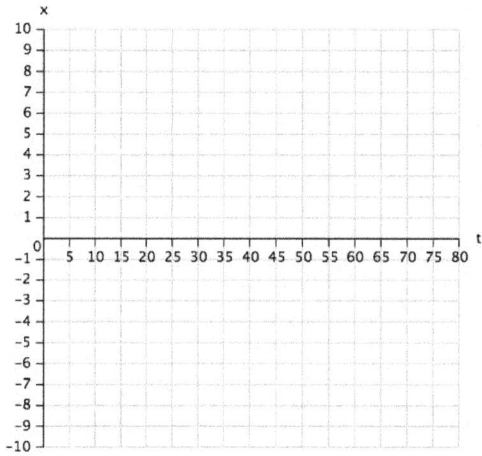

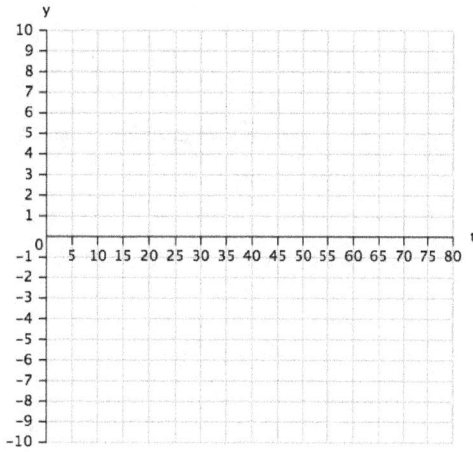

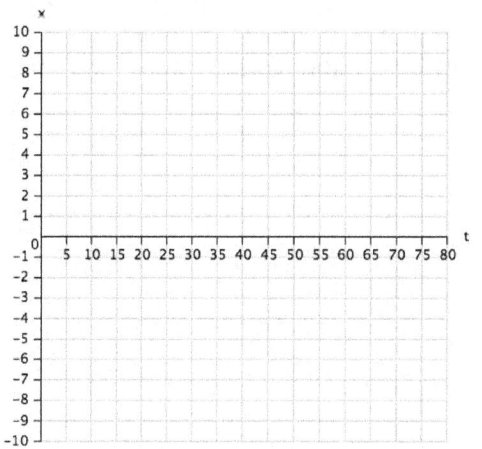

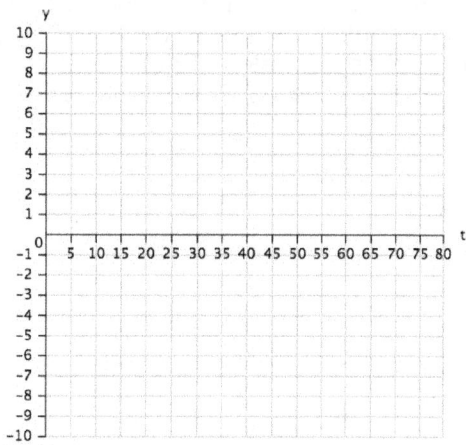

x

5 10 15 20 25 30 35 40 45 50 55 60 65 70 75 80 t

y

5 10 15 20 25 30 35 40 45 50 55 60 65 70 75 80 t

Player One Doodle:

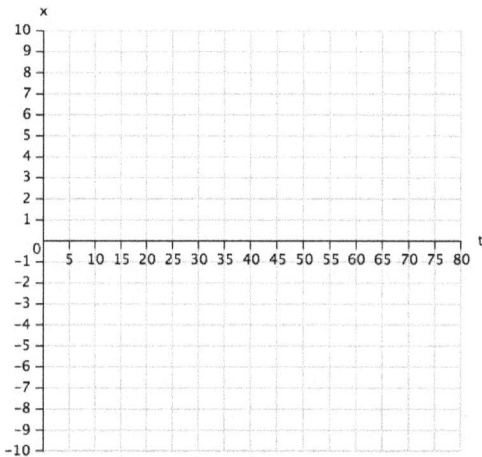

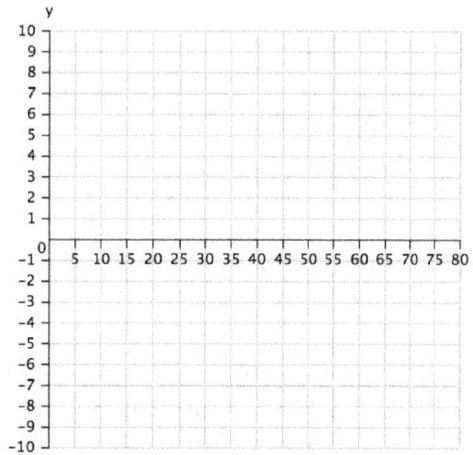

x

5 10 15 20 25 30 35 40 45 50 55 60 65 70 75 80 t

y

5 10 15 20 25 30 35 40 45 50 55 60 65 70 75 80 t

Player Two Doodle:

x

5 10 15 20 25 30 35 40 45 50 55 60 65 70 75 80 t

y

5 10 15 20 25 30 35 40 45 50 55 60 65 70 75 80 t

Teaching Guide

The Doodle Duel

App: Action Grapher

Overview:

Students take turns drawing a doodle, recording the x vs. arc length and y vs. arc length, and then challenging the other student to reproduce their doodle from just the x vs. arc length and y vs. arc length graphs.

Common Core State Standards:

- CCSS.Math.Content.8.EE.B.5
- CCSS.Math.Content.HSF.IF.B.4
- CCSS.Math.Content.HSF.IF.B.6

Encyclopedia of Algebraic Thinking:

- Analysis of Change: Interpreting Graphs
- Analysis of Change: Graphs as a Literal Picture

Description:

In this activity students are given the freedom to create their own challenges for each other. Students will need to first familiarize themselves with the Doodle section of the app Action Grapher. The way in which this section of the app takes a two-dimensional doodle and represents that doodle on an x vs. arc length graph and a y vs. arc length graph may be initially confusing to students. For that reason this Doodle Duel activity includes a length limit that the teacher or students must agree upon. You may also want to experiment with setting other limits on the doodles that students draw - such as only drawing straight lines or only drawing certain shapes. By applying the appropriate restrictions on a student-by-student basis, you will be able to provide the right level of difficulty for each student.

Extensions:

- Doodle Pictionary
- The Doodle Challenge

Doodle Pictionary

For use with the iPad app *Action Grapher*

Name: Date: Period:

If this activity, your instructor will present for you an x vs. arc length and y vs. arc length graph. Your task is to determine the object that results from the doodle that produces those two graphs. Use this sheet to record the object for each round:

Round 1: _____   Round 2: _____

Round 3: _____   Round 4: _____

Round 5: _____   Round 6: _____

Round 7: _____   Round 8: _____

Round 9: _____   Round 10: _____

Round 11: _____   Round 12: _____

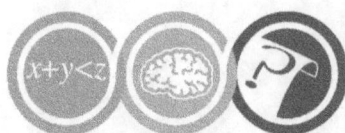

Teaching Guide

Doodle Pictionary

App: Action Grapher

Overview:

Students play a variation of the game Pictionary as they practice transitioning from x vs. arc length and y vs. arc length graphs to a complete two dimensional doodle.

Common Core State Standards:
- CCSS.Math.Content.8.EE.B.5
- CCSS.Math.Content.HSF.IF.B.4
- CCSS.Math.Content.HSF.IF.B.6

Encyclopedia of Algebraic Thinking:
- Analysis of Change: Interpreting Graphs
- Analysis of Change: Graphs as a Literal Picture

Description:

This activity adds some playfulness to the process of moving between different graphical representations of the same information. Students should be familiar with the basic process by which the doodle is created from the x vs. t and y vs. t graphs prior to participating in this activity.

Teachers are encouraged to create their own objects, sketch (or screen shot) the x vs. t and y vs. t graphs, and then use those graphs to present to students in this game. To get you started, see the following example graphs that fall within the topic 'shapes':

- Doodle Pictionary: Shapes

Key: 1) Square; 2) Triangle; 3) Rectangle; 4) Circle; 5) Diamond; 6) Ellipse; 7) Parallelogram; 8) Pentagon.

Extensions:

Let students challenge each other by creating their own doodles of objects and then swapping the x vs. t and y vs. t graphs and see if they can identify the original object.

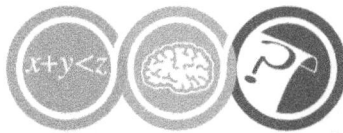

Center for Algebraic Thinking
Teaching Resource

Doodle Pictionary: Shapes

App: Action Grapher

Present each set of x vs. arc length and y vs. arc length graphs to students and challenge them to identify the shape that results from the doodle that produces those graphs.

1.

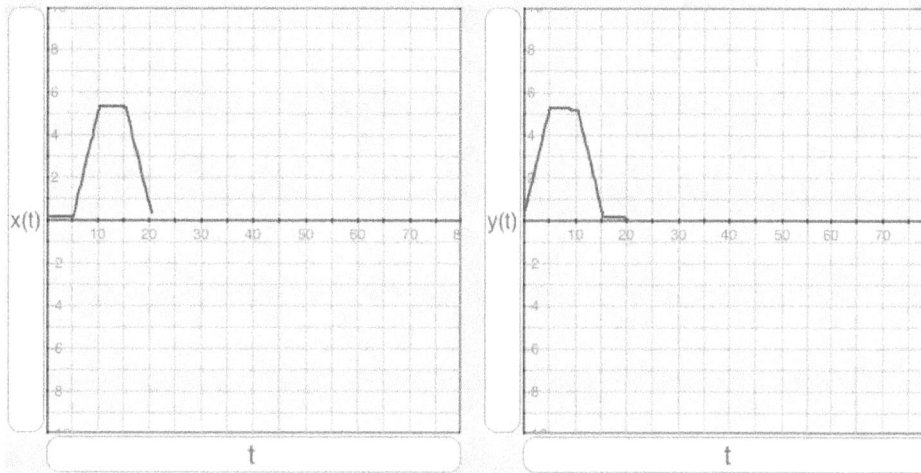

2.

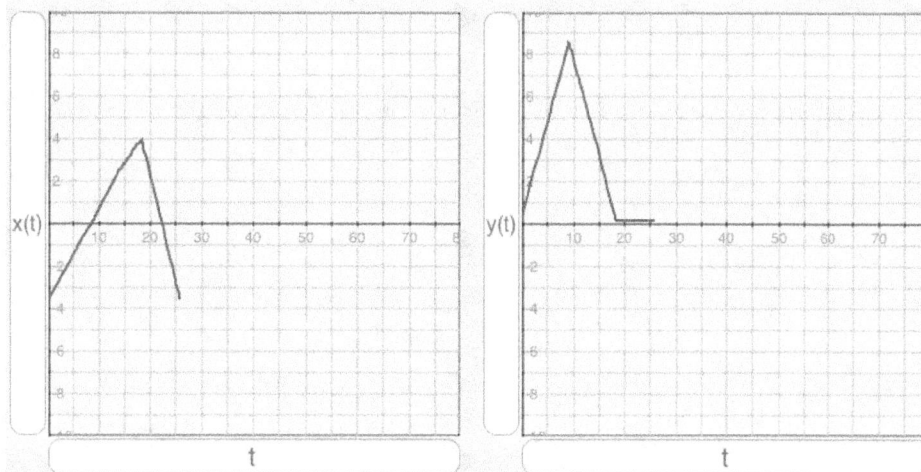

3.

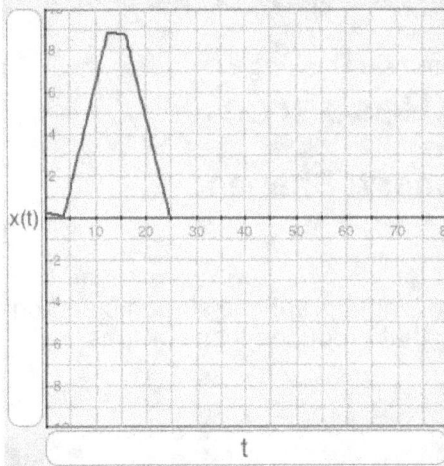

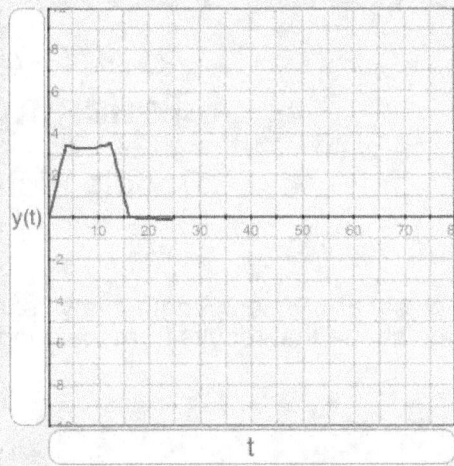

4.

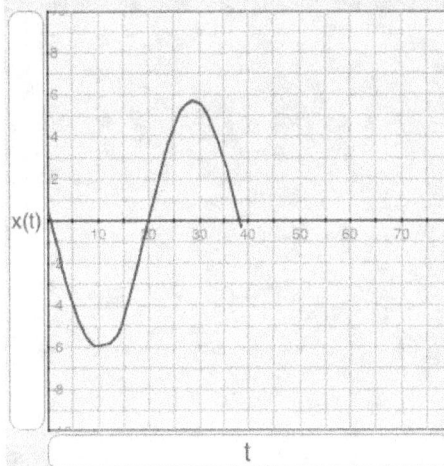

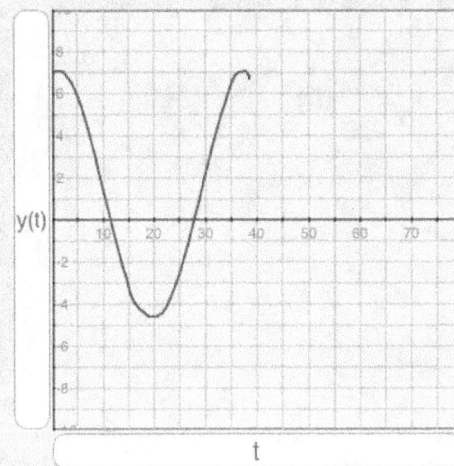

5.

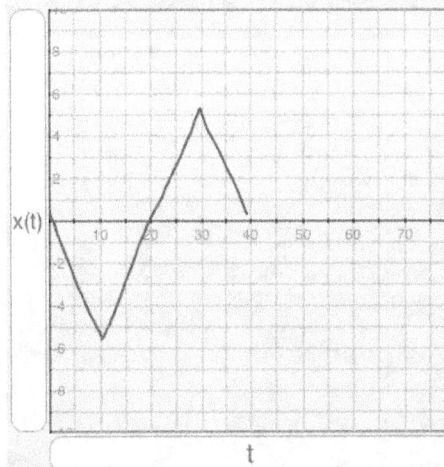

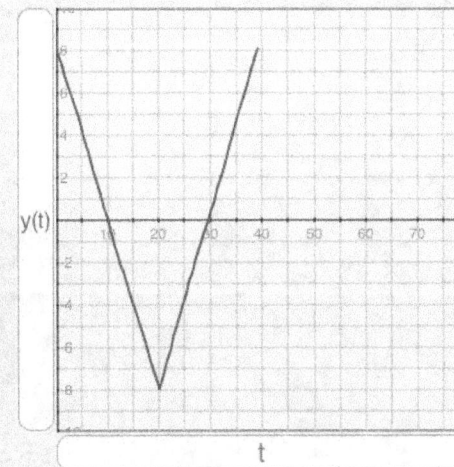

6.

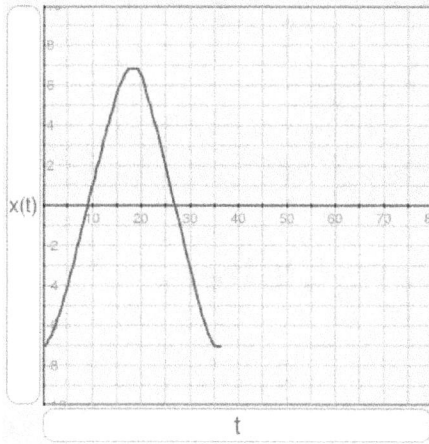

x(t)

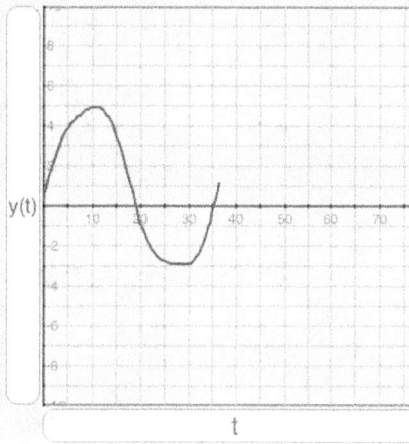

y(t)

7.

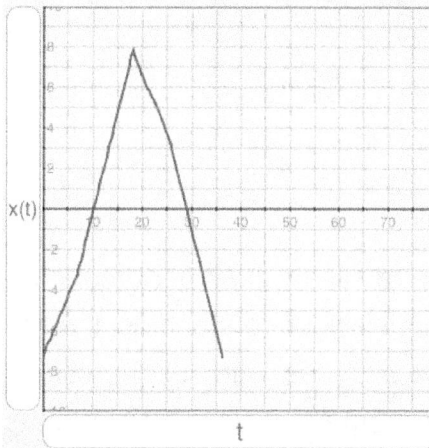

x(t)

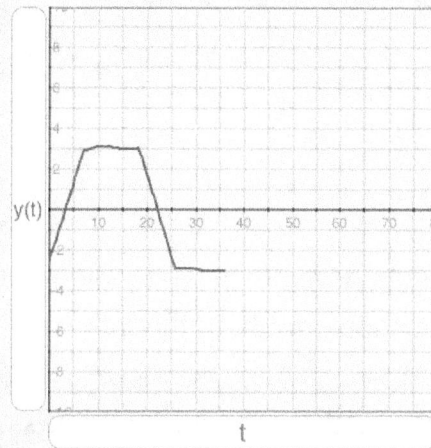

y(t)

8.

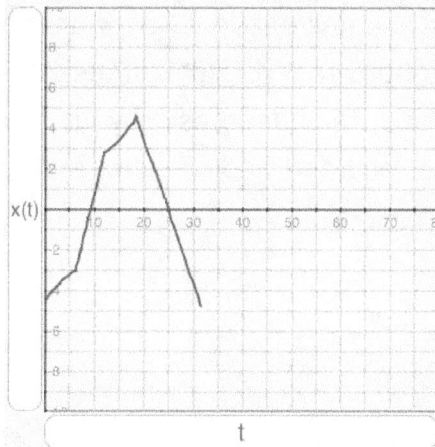

x(t)

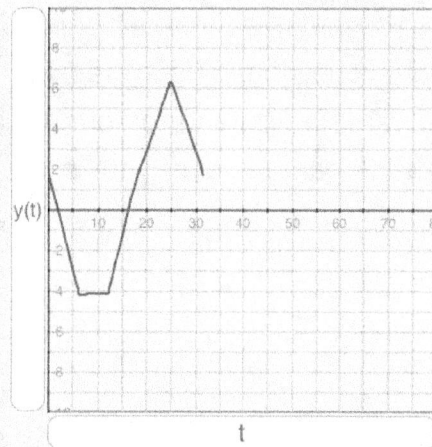

y(t)

Algebra Card Clutter

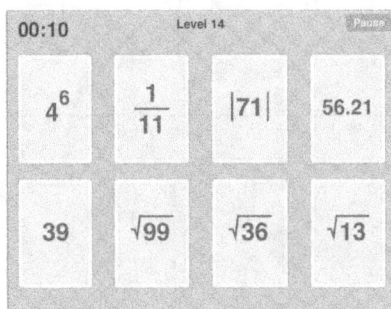

00:10   Level 14   Pause

4^6	$\dfrac{1}{11}$	$|71|$	56.21
39	$\sqrt{99}$	$\sqrt{36}$	$\sqrt{13}$

Help students develop understanding of the relative values of numeric expressions. We know from research that students have difficulty understanding the relative values of integers. In this game, you have a bunch of cards with numbers and other numeric expressions, and your pesky friend has put them in the wrong order on a table. Your job is to pick up the cards in the order, so you can put them away. Each level is one table full of cards, and later levels introduce more complicated expressions. Levels either have 5 or 10 cards, so the difficulty can be increased without transition to more complicated expressions. Each session keeps track of total time. Mistakes do not remove the card, but do add to your total time, thus decreasing your "score" (i.e. your time). When the game is over your score is recorded and you can enter your name in the leader boards.

Teaching Guide

Two-Player Card Cutter

App: Algebra Card Clutter

Overview:
In this activity students cooperatively play the game *Algebra Card Clutter* in which they must select the lowest card on the screen. The numbers on the cards are given in a variety of formats including positive and negative integers, decimals, fractions, absolute values, square roots, and exponentials.

Common Core State Standards:
- CCSS.Math.Content.6.NS.C.6
- CCSS.Math.Content.6.NS.C.7
- CCSS.Math.Content.8.NS.A.2

Encyclopedia of Algebraic Thinking:
- Algebraic Relations: Negative Numbers
- Algebraic Relations: Rational Numbers

Description:
The purpose of this resource is to turn the engaging and challenging app *Algebra Card Clutter* into a two-player game in order to tap into the benefits of social learning and provide a structured environment for students to help each other when they run into problems that they cannot solve individually.

In this activity students alternate turns, with a key factor being that on each turn the student must make the first attempt on their own. Only if the student gets that first attempt wrong can the second student jump in to help. The idea is that those initial errors will lead to a discussion about the correct answer, and that together (or at times with the instructor's help as well) they will be able to figure out how to move forward and fill in the gaps that made them unable to complete the challenge on their own. Since the students are recording the levels they have trouble with, and explaining the troubles they had, they will be more likely to remember what they learned from each mistake. Teachers can also use the record to get a quick feel for what students are struggling with the most.

Extensions:
- Card Clutter Progress Monitor
- Student-Created Card Clutter

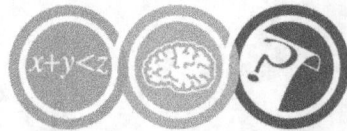

Two-Player Card Clutter

For use with the iPad app *Algebra Card Clutter*

Names: Date: Period:

In this activity you and a partner will cooperatively play the game Algebra Card Clutter *on the iPad.*

Getting Started

From the home screen of the app, tap the 'Start' button at the top of the menu. Then select the mode you would like to play. If you have not played this game before, tap 'Full Course'. If you have played it before and feel ready for more of a challenge, tap 'Advanced Course'.

Gameplay

Once the game starts, you and your partner will take turns selecting the lowest card on the screen. When it is your turn, your first attempt has to be on your own (without help from your partner). But if you select the wrong card on your first try, then you and your parter can work together to select the correct card. Continue alternating turns throughout the game.

If the time ever runs out on you, then use the space below to record the correct solution for that level, and explain the trouble you had completing the level. Then tap 'Retry' and continue playing. Follow that process until you have completed all 14 levels, and then record you final time at the bottom on the next page.

Level: Solution:

Explanation of the trouble you had with this level:

Level: Solution:

Explanation of the trouble you had with this level:

Level: Solution:

Explanation of the trouble you had with this level:

Level: Solution:

Explanation of the trouble you had with this level:

Continue playing until you finish all 14 levels, and then record your final time below.

Final Time: _____

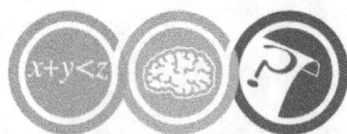

Center for Algebraic Thinking

Card Clutter Progress Monitor

For use with the iPad app *Algebra Card Clutter*

Name: Date: Period:

In this activity you will record your progress as your work through the game Algebra Card Cutter

Getting Started

From the home screen of the app, tap the 'Start' button at the top of the menu. Then select the mode you would like to play. If you have not played this game before, tap 'Full Course.' If you have played it before and feel ready for more of a challenge, tap 'Advanced Course.'

Gameplay

Your task in this game is to tap the cards on the screen from the smallest card value to the largest. Each time you make an error some time will be added to your score. If the time ever runs out on you, then use the space below to record the correct solution for that level, and explain the trouble you had completing it. Then tap 'Retry' and continue playing. Follow that process until you have completed all 14 levels, and then record you final time at the bottom on the next page.

Level: Solution:

Explanation of the trouble you had with this level:

Level: Solution:

Explanation of the trouble you had with this level:

Level: Solution:

Explanation of the trouble you had with this level:

Level: Solution:

Explanation of the trouble you had with this level:

Level: Solution:

Explanation of the trouble you had with this level:

Continue playing until you finish all 14 levels, and then record your final time below.

Final Time: _____

Teaching Guide

Card Clutter Progress Monitor

App: Algebra Card Clutter

Overview:

In this activity students work through the game *Algebra Card Clutter* in which they must tap cards from the smallest value to the largest in progressively harder levels. With this paper resource students record and reflect on any situations in which they are not able to pass a level on their own.

Common Core State Standards:
- CCSS.Math.Content.6.NS.C.6
- CCSS.Math.Content.6.NS.C.7
- CCSS.Math.Content.8.NS.A.2

Encyclopedia of Algebraic Thinking:
- Algebraic Relations: Negative Numbers
- Algebraic Relations: Rational Numbers

Description:

The strength of this app is that the difficulty builds in a way that makes it accessible to students with varying degrees of prior knowledge. The form of the numbers on the cards progress from integers, to fractions, to decimals, to absolute values, to exponents, and finally to square roots. Once a student runs into a topic that is difficult enough that they cannot pass the level, then this resource provides the space to stop and reflect on that topic, and get additional help if needed. They record the correct order of the cards (which is presented to them within the app) and then they write about the trouble they had with that level. This reflection will both help them to work through confusions on their own as well as solidify the new knowledge.

Extensions:

- Two-Player Card Clutter
- Student-Created Card Clutter

Center for Algebraic Thinking

Student-Created Card Clutter

For use with the iPad app *Algebra Card Clutter*

Name: Date: Period:

In the iPad app Algebra Card Clutter *your task is to sort cards from the smallest number to the largest number. In this activity you and your classmates will* **create your own cards**, *pool those cards together, and then play the same game hands-on.*

Step 1: Card Creation
The first step is to create the physical cards that you and your classmates will use to play the game. You need to **create a total of 12 cards, two in each of the categories given below**. At least five of your cards need to have negative values. Here are the categories:

Integers	Decimals	Exponentials
Fractions	Absolute Values	Square Roots

So, create 12 cards total, two from each category above, and remember that at least five of your cards need to have negative values.

Step 2: Combining Cards
Now you will combine the cards you created with the cards of your classmates. The idea here is to create a large stack of cards that your group will be able to use to play the game. So pool all of your cards together, either all of those within your table group, or all of those within your entire class (depending on the directions from your instructor).

Step 3: Play the Game!
After mixing all of the cards together, and then redistributing them to the different groups, you will starting playing the game. In groups of two, start with your stack of cards face down on the table in front of you. You and your partner will take turns. One player starts by selecting the first five cards from the deck and placing them face up on the desk. That player then has a maximum time of **one minute** to order those cards from the smallest value to the largest value. Once that player is satisfied with the order (or after a minute has passed) you will record *the decimal values* of each of those cards on the following page in order to see if that player ordered the cards correctly. You can use a calculator for this step, but you cannot use a calculator when doing the ordering. After recording the decimal values, check the appropriate blank to show whether or not the order was correct. The second player then follows the same process. Continue going back and forth and recording the results. If you run out of cards, you can reshuffle your deck and continue playing.

Round One

Player 1: _____ _____ _____ _____ _____   Correct Order? Y ____ N ____

Player 2: _____ _____ _____ _____ _____   Correct Order? Y ____ N ____

Round Two

Player 1: _____ _____ _____ _____ _____   Correct Order? Y ____ N ____

Player 2: _____ _____ _____ _____ _____   Correct Order? Y ____ N ____

Round Three

Player 1: _____ _____ _____ _____ _____   Correct Order? Y ____ N ____

Player 2: _____ _____ _____ _____ _____   Correct Order? Y ____ N ____

Round Four

Player 1: _____ _____ _____ _____ _____   Correct Order? Y ____ N ____

Player 2: _____ _____ _____ _____ _____   Correct Order? Y ____ N ____

Round Five

Player 1: _____ _____ _____ _____ _____   Correct Order? Y ____ N ____

Player 2: _____ _____ _____ _____ _____   Correct Order? Y ____ N ____

Round Six

Player 1: _____ _____ _____ _____ _____   Correct Order? Y ____ N ____

Player 2: _____ _____ _____ _____ _____   Correct Order? Y ____ N ____

Round Seven

Player 1: _____ _____ _____ _____ _____   Correct Order? Y ____ N ____

Player 2: _____ _____ _____ _____ _____   Correct Order? Y ____ N ____

Round Eight

Player 1: _____ _____ _____ _____ _____   Correct Order? Y ____ N ____

Player 2: _____ _____ _____ _____ _____   Correct Order? Y ____ N ____

Center for Algebraic Thinking

Teaching Guide

Student-Created Card Clutter

App: Algebra Card Clutter

Overview:
In this activity students bring the game Algebra Card Clutter into the real world by making their own physical cards and then playing the game hands-on with classmates.

Common Core State Standards:
- CCSS.Math.Content.6.NS.C.6
- CCSS.Math.Content.6.NS.C.7
- CCSS.Math.Content.8.NS.A.2

Encyclopedia of Algebraic Thinking:
- Algebraic Relations: Negative Numbers
- Algebraic Relations: Rational Numbers

Description:
After bring introduced to the idea of this game via the iPad app *Algebra Card Clutter*, this activity will bring it into the real world and give students ownership of it.

You begin by having each student create their own cards. Each card has just one number on it, and students must create a total of 12, with two numbers in each of the following categories: integers, fractions, decimals, absolute values, exponentials, and square roots. By requiring that students create two cards for each category, it forces students to write numbers in forms they might not normally use. Depending on students' prior knowledge, teachers can adjust this step by including only those categories with which students are comfortable. To ensure that students explore the entire number line, a second requirement is that at least five of the cards contain negative values.

You then pool all of the cards together, mix them, and then redistribute them to groups of two. This mixing and redistributing means that students will get many cards that they did not create, but were instead created by classmates, which adds some variability into the values that students will be playing with.

In groups of two, students then use their stack of cards to play the game, in which players alternate selecting five cards from the deck and then ordering those cards from the smallest value to the largest. To check their ordering they convert each value to a decimal and record those decimals on a separate piece of paper. This record provides a means for accountability - the teacher can collect this at the end of the activity.

Extensions:
You can extend this activity by varying the categories used (leaning toward either the easier categories or more difficult ones). You can also vary the number of cards created by each student, and the number of cards selected from the deck for each round of play.

Algebra Equation Builder

Research tells us that students struggle to understand the nature of the equal sign, an essential idea in algebra. Rather than seeing the equal sign as a message to compute from left to right, seeing it as a sense of balance. *Algebra Equation Builder* encourages students to develop that sense of balance as they build and decompose equations. Four levels of equation building are provided.

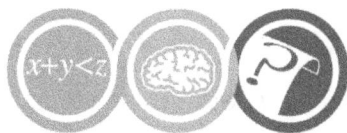

Center for Algebraic Thinking

Algebra Equation Builder Card Game

For use with the iPad app *Algebra Equation Builder*

Name: Date: Period:

In this activity you will build an equation using cards on which is written either a number, an operation, or an equal sign. With each set of cards, you must arrange the cards in a way that results in a true equation. You know which cards are part of the same set by looking for the same letter in the lower right corner of the card.

Example

Let's say that you were given the following set of cards:

-6	8	8	10	-	=	+
A	A	A	A	A	A	A

Here is how you could arrange those cards in a way that would create a true equation:

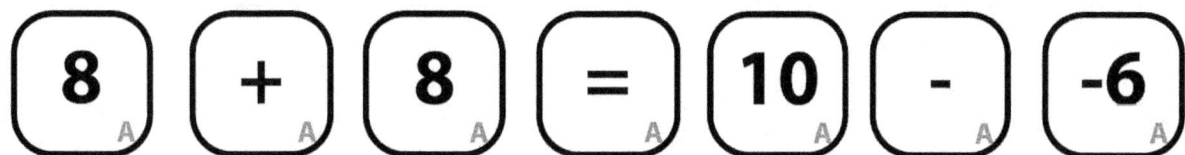

8	+	8	=	10	-	-6
A	A	A	A	A	A	A

Activity Log

Use the space below to record which card sets you solve, and the equation that you create with each set.

1. Set Letter: _____   Equation: _____

2. Set Letter: _____   Equation: _____

3. Set Letter: _____   Equation: _____

4. Set Letter: _____   Equation: _____

5. Set Letter: _____   Equation: _____

6. Set Letter: _____   Equation: _____

7. Set Letter: _____   Equation: _____

8. Set Letter: _____   Equation: _____

9. Set Letter: _____   Equation: _____

10. Set Letter: _____   Equation: _____

Center for Algebraic Thinking

Teaching Guide

Algebra Equation Builder Card Game

App: Algebra Equation Builder

Overview:

In this activity students play an analog version of the challenge they will face in the app *Algebra Equation Builder*. This activity can be used as an introduction to that app or as an activity for students without iPads to play while others with iPads are using the app.

Common Core State Standards:
- CCSS.Math.Content.6.EE.A.4
- CCSS.Math.Content.7.NS.A.1
- CCSS.Math.Content.7.EE.A.1
- CCSS.Math.Content.HSA.SSE.B.3
- CCSS.Math.Content.HSA.CED.A.1

Encyclopedia of Algebraic Thinking:
- Algebraic Relations: Negatives
- Variable and Expressions: Letter used as a generalized number

Description:

In this activity students are given a set of cards that contain either a number, an operation, or an equal sign. The challenge is to arrange those cards in a way that creates a true equation. To produce the card sets that students will use, instructors can either print and cut out the cards on the following pages, or use index cards to create their own set. The students can use the activity log on the previous page to keep track of which sets they have solved and their solution for each set. Instructors can use that log as an accountability check and as a way to see if solutions are correct.

Extensions:
- Algebra Equation Builder Progress Monitor

-9 A	-5 A	-4 A
1 A	· A	= A
- A	-5 B	-5 B
-5 B	2 B	= B

· **B**	= **B**	-6 **C**
-3 **C**	1 **C**	3 **C**
· **C**	= **C**	- **C**
-12 **D**	-11 **D**	-10 **D**

-9 D

- D

= D

- D

-4 E

6 E

10 E

12 E

- E

= E

+ E

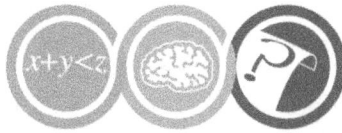

Center for Algebraic Thinking

Algebra Equation Builder Progress Monitor

For use with the iPad app *Algebra Equation Builder*

Name: Date: Period:

Use the space below to record your progress as you advance through the levels of the app *Algebra Equation Builder*. Start with Level 1. For each level, record how long it takes you to make it through the level. If you get stuck on an equation, record the equation in the space provided below, then restart that level and give it another go. Once you make it through a level, move on to the next!

Level 1
Time taken to complete level: _____

Equations that gave you trouble (if needed):

_____

_____

_____

Level 2
Time taken to complete level: _____

Equations that gave you trouble (if needed):

_____

_____

_____

Level 3

Time taken to complete level: _____

Equations that gave you trouble (if needed):

_____

_____

_____

Level 4

Time taken to complete level: _____

Equations that gave you trouble (if needed):

_____

_____

_____

Teaching Guide

Algebra Equation Builder Progress Monitor

App: Algebra Equation Builder

Overview:

In this activity students record their progress (and also any struggles) as they work through the levels of the app *Algebra Equation Builder*.

Common Core State Standards:
- CCSS.Math.Content.6.EE.A.4
- CCSS.Math.Content.7.NS.A.1
- CCSS.Math.Content.7.EE.A.1
- CCSS.Math.Content.HSA.SSE.B.3
- CCSS.Math.Content.HSA.CED.A.1

Encyclopedia of Algebraic Thinking:
- Algebraic Relations: Negatives
- Variable and Expressions: Letter used as a generalized number

Description:

The purpose of this activity is to provide space for students to track their progress through the increasingly-difficult levels of the app *Algebra Equation Builder*. If they encounter a problem they cannot solve, they have space to record that problem on the page (which might trigger a solution that they hadn't noticed within the app). Even if they don't find a solution to that particular problem, they are encouraged to restart that level and continue working. The difficulty of the levels increases steadily, so even students that are grade-level proficient will find challenges within the higher levels.

Extensions:

If students are doing particularly well with the problems presented in this app, you could encourage them to create their own problems using index cards (similar to the activity Algebra Equation Builder Card Game), and then share the problems they have created with other students. You could also encourage students to see if they can beat their best time for any given level.

Compound Interest Simulation

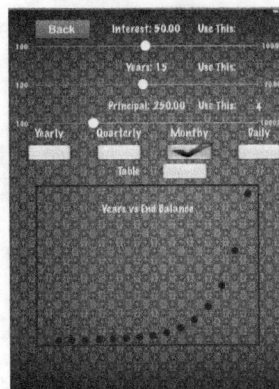

The Compound Interest Simulator helps students learn about the impact that changes in interest rates, time, and principal have upon the ending balance. Using sliders to adjust the different variables, students can watch how a table or graph are impacted by their manipulation. Students also have the option of considering interest compounded yearly, quarterly, monthly, or daily.

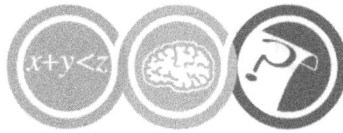

Center for Algebraic Thinking
Funding Your Dreams

For use with the iPad app
Compound Interest Simulation

Name: Date: Period:

In this activity you will have $1,000 to invest today in order to fund your dreams for the future. But which dreams will you be able to afford? Let's find out.

First, let's figure out what dreams you're going to try to fund. After writing down your current age below, write down your age in the time increments listed below. Next to those ages write down a dream activity - a **reasonable** vacation or free-time activity - and the estimated cost of that activity.

Current Age:

	Age	Dream Activity	Estimated Cost
In 5 Years			
In 10 Years			
In 15 Years			
In 20 Years			
In 25 Years			
In 30 Years			

Next we'll see if you'll be able to afford any of those activities by investing $1,000 today. Continue on to the next page to find out.

The next step is to calculate how much return you'll get on $1,000 invested today. Because interest rates tend to be larger the longer you invest your money, you'll see an increase in the interest rate as you increase the total amount of time your money is invested. Fill in the chart below to determine if any of your dream activities can be funded.

Principal	# of Years	Interest Rate	Final Balance	Dream Funded? Y/N
$1,000	5	4%		
$1,000	10	5%		
$1,000	15	6%		
$1,000	20	7%		
$1,000	25	8%		
$1,000	30	9%		

Based on your calculations, if you had to pick just *one* of those investment options (you can't pick them all - you've only got *one* chunk of $1,000 to invest) which would you choose, and why?

BONUS: Were there any activities that you *really* want to do, but that cost too much? How much would you have to invest today to fully fund that dream?

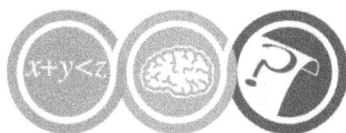

Center for Algebraic Thinking

Teaching Guide

Funding Your Dreams

App: Compound Interest Simulation

Overview:

In this activity students will plan dream vacation activities at different points in their life, estimate how much each of those activities will cost, and then determine if a $1,000 investment made today will fund any of those dreams.

Common Core State Standards:
- CCSS.Math.Content.HSF.LE.A.3
- CCSS.Math.Content.HSF.LE.B.5

Encyclopedia of Algebraic Thinking:
- Analysis of Change: Interpreting Graphs
- Analysis of Change: Scaling

Description:

This activity makes the idea of compound interest relevant by giving students the opportunity to think about what would happen if they invested $1,000 today. The activity is designed to be completed individually, but an introduction will be necessary if students aren't yet familiar with the idea of principal, interest rate, and compound interest.

Once students have had a chance to complete the activity, you could have a class discussion about which dreams were funded, which were not, and your students' overall impressions about investment after seeing real numbers attached to things that students are likely emotionally invested in.

Extensions:

- The Compound Interest Estimation Game
- How Often To Compound?

Center for Algebraic Thinking

How Often To Compound?

For use with the iPad app
Compound Interest Simulation

Name: Date: Period:

In this activity you will look at the effects of different rates of compounding interest on an investment.

1. When making an investment, which is best for you as the investor - to compound interest yearly, quarterly, monthly, or daily? Why?

Let's test your response to the question above. You're going to have $1,000 to invest at an interest rate somewhere between 5% and 15%, and then you'll track how much interest is earned each year on that investment for each of the compounding options listed above.

2. First, pick an interest rate (your choice) between 5% and 15%.

   Principal: $1,000   Interest Rate: _____   Number of Years: 10

3. Next, record the *interest earned* after each year for each of the compounding options.

Year	Interest Earned Compounded Yearly	Interest Earned Compounded Quarterly		
1				
2				
3				
4				
5				
6				
7				
8				
9				
10				

4. Based on your calculations entered in the table on the previous page, was your response to Question 1 correct?

5. How much of a difference was there in the amount of interest earned when it was compounded yearly compared to when it was compounded daily?

6. In what ways would your results be different if you chose a different interest rate?

7. In what ways would your results be the same if you chose a different interest rate?

Center for Algebraic Thinking

Teaching Guide

How Often To Compound?

App: Compound Interest Simulation

Overview:

In this activity students will get a feel for the differences in the amount of interest earned when that interest is compounded yearly, quarterly, monthly, or daily.

Common Core State Standards:

- CCSS.Math.Content.HSF.LE.A.3
- CCSS.Math.Content.HSF.LE.B.5

Encyclopedia of Algebraic Thinking:

- Analysis of Change: Interpreting Graphs
- Analysis of Change: Scaling

Description:

Though the initial question of *which compounding option is best for the investor* is a relatively simple one, this activity will allow students to discover first hand the effect of that decision on an investment over time.

Since each individual student will chose the interest rate, at the end of the activity you will have a variety of data that you can discuss as a class. You could have individual students share their results and then discuss their responses to questions 6 and 7 regarding what would be the same and what would be different if you had chosen a different interest rate. That will give students the opportunity to check their reasoning against actual results.

Extensions:

- The Compound Interest Estimation Game
- Funding Your Dreams

Center for Algebraic Thinking

The Compound Interest Estimation Game

For use with the iPad app
Compound Interest Simulation

Name: Date: Period:

In this activity we're going to play an estimation game - you'll be estimating how much money you would make from the interest on an investment after certain periods of time, starting principal, and interest rate. Your instructor will give you the starting conditions for each round.

Principal	Interest Rate	# of Years	Estimated Final Balance	Actual Final Balance

Look back at how your estimated final balances compare to the actual final balance of the investments above. What have you learned about compound interest investments?

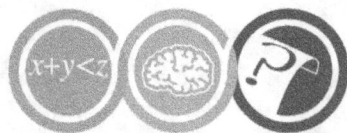

Center for Algebraic Thinking

Teaching Guide

The Compound Interest Estimation Game

App: Compound Interest Simulation

Overview:
In this activity students will get a feel for the numbers surrounding compound interest through an estimation game in which you will provide the starting conditions and students will estimate the final balance.

Common Core State Standards:
- CCSS.Math.Content.HSF.LE.A.3
- CCSS.Math.Content.HSF.LE.B.5

Encyclopedia of Algebraic Thinking:
- Analysis of Change: Interpreting Graphs
- Analysis of Change: Scaling

Description:
Begin by explaining any new vocabulary and asking students to imagine that they have a bit of money to invest for the future. Hand each student a copy of The Compound Interest Estimation Game worksheet and then complete the following steps for each round:

1. Announce the starting conditions: the principal, interest rate, and number of years the money will be invested.
2. Ask students to individually write their estimates of what the final balance will be.
3. Request a handful of estimates from students to write on the board (or systematically pick students so that in the end each student will have provided a public estimate).
4. Use the Compound Interest Simulation app to calculate the final balance, and reveal that balance to the class.
5. (Optional) Give a small prize to the student that made the closest public estimate.

Repeat the steps above, adjusting just one variable at a time, and discuss how that change in variable affected the result. Once you have adjusted each variable, start again with vastly different initial conditions. After completing many rounds, give students time to individually answer the question below the table where they have been recording their estimates.

Extensions:
- Funding Your Dreams
- How Often To Compound?

Cover Up Math

$$\frac{48}{(x-3)^2} + 17 = 29$$

Cover up this expression ⌄

⌃ Home   Solve this equation ＞

In math classes we tend to ignore students' intuition when helping them learn how to solve equations. Sometimes, their intuition is more likely to help them find meaning than rote procedures. Young students can solve complex equations when they don't have to rely on a mechanical use of the order of operations. Cover Up helps students develop a strategy for solving algebraic equations that is more intuitive than the traditional use of order of operations. Given an equation that is "messy" with fractions, exponents, square roots, etc., students cover up the challenging part of the equation to make it more intuitive to solve. Five different levels of problems are included. A great way to develop confidence in solving challenging algebraic equations in a way that makes sense!

69

Center for Algebraic Thinking

Sticky Note Cover Up Math

For use with the iPad app *Cover Up*

Name: Date: Period:

In this activity you will learn how to use sticky notes to cover up certain parts of an equation so that you can focus in on the solution one step at a time. After learning the process you will get to practice your new skills with the iPad app Cover Up.

Here's how the process works. Let's say you are given the equation **7 = 6x - 11** and your task is to solve for x. You can use sticky notes to help you in the following way:

$7 = 6x - 11$ ① Write equation

sticky note →

$7 = \boxed{} - 11$ ② Cover up part of equation with sticky note

$7 = \boxed{18} - 11$ ③ Determine what number is needed, write it on sticky note

$7 = 6x - 11$
$6x = \boxed{18}$ ④ Remove sticky note, write new equation below old

$x = 3$ ⑤ Solve for x

$7 = 6x - 11$
$7 = 6(3) - 11$
$7 = 18 - 11$
$7 = 7 \checkmark$ ⑥ Check your work

Using the cover up method allows you to get started on a problem by ignoring the complicated part of an equation - the part with the variable. If you look at the equation as it is written in Step 1, you might not know where to start. But once you cover up the **6x** with a sticky note, it looks a whole lot more approachable. Now you just have to figure out what number minus 11 equals 7. A little mental math tells you that number is 18. So now you know that whatever is behind the sticky note has to be equal to 18. The last step is to figure out what number times 6 equals 18. A bit more mental math will tell you that the answer is 3. A final check of your work confirms that your answer is correct.

Use the space below to give this method a try with the following equation: **15 = 4x + 3**

Center for Algebraic Thinking

Teaching Guide

Sticky Note Cover Up Math

App: Cover Up

Overview:

In this activity the student is introduced to the Cover Up Math Method with a detailed example and full explanation, and then given a chance to practice the method on a new problem.

Common Core State Standards:

- CCSS.Math.Content.6.EE.A.2
- CCSS.Math.Content.6.EE.B.5
- CCSS.Math.Content.6.EE.B.6
- CCSS.Math.Content.6.EE.B.7

- CCSS.Math.Content.7.EE.B.3
- CCSS.Math.Content.8.EE.C.7
- CCSS.Math.Content.HSA.SSE.A.1
- CCSS.Math.Content.HSA.REI.A.1

Encyclopedia of Algebraic Thinking:
- Algebraic Relations: Equations Involving Negatives
- Algebraic Relations: Flexible Use of Solution Strategies
- Algebraic Relations: One and two step equations
- Algebraic Relations: Student Intuition and Informal Procedures
- Variables and Expressions: Letter Used as a Specific Unknown

Description:
This activity is designed as an introduction to the Cover Up Math Method. The purpose is to explain the process of the method prior to letting students practice the method with the iPad app *Cover Up*. If students jump right into using the app without an explanation of the method, they will be less likely to transfer the skills when working on problems on their own. One approach to this activity is to have students work through it individually. Another option is to hand out the activity and then work through the example as a class, using an actual sticky note on the board or projected from a doc cam. By introducing this method with pen and paper first, and then following up with digital practice (see the extension below), students will be better prepared to use it when they return to pen and paper work.

Extensions:
- Cover Up Math App Progress Monitor
- Sticky Note Cover Up Challenge
- Comparison Of Methods: Cover Up VS Algorithmic
- The Mental Math Cover Up Challenge

Center for Algebraic Thinking

Cover Up Math App Progress Monitor

For use with the iPad app *Cover Up*

Name: Date: Period:

Use this sheet to record the results obtained using the Cover Up *app and to keep track of the levels and problems that you have completed.*

Level	Equation	Solution

Center for Algebraic Thinking

Teaching Guide

Cover Up Math App Progress Monitor

App: Cover Up

Overview:

This resource allows students to keep track of their progress as they work through the various levels of the app *Cover Up*.

Common Core State Standards:
- CCSS.Math.Content.6.EE.A.2
- CCSS.Math.Content.6.EE.B.5
- CCSS.Math.Content.6.EE.B.6
- CCSS.Math.Content.6.EE.B.7
- CCSS.Math.Content.7.EE.B.3
- CCSS.Math.Content.8.EE.C.7
- CCSS.Math.Content.HSA.SSE.A.1
- CCSS.Math.Content.HSA.REI.A.1

Encyclopedia of Algebraic Thinking:
- Algebraic Relations: Equations Involving Negatives
- Algebraic Relations: Flexible Use of Solution Strategies
- Algebraic Relations: One and two step equations
- Algebraic Relations: Student Intuition and Informal Procedures
- Variables and Expressions: Letter Used as a Specific Unknown

Description:
This resource serves two purposes. First, it provides students the space to monitor their own progress through the app while simultaneously bridging the gap between digital tools and pencil and paper work. Second, it provides teachers with a record of the work that students have completed related to the app. By including sections for the student to record each problem as well as the solution, it helps to ensure that each student is actually completing the problems put forward by the app.

Extensions:
- Sticky Note Cover Up Challenge
- Comparison of Methods: Cover Up VS Algorithmic
- The Mental Math Cover Up Challenge

73

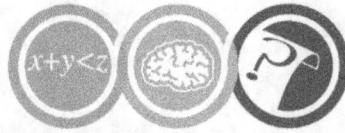

Sticky Note Cover Up Challenge

For use with the iPad app *Cover Up*

On a separate piece of paper, use the cover up math method to solve the equations below. If you need a refresher, check out the app Cover Up*:*

Level 1:

1) $5x + 7 = 32$

2) $12 - 3x = 6$

3) $9 = 25 - 4x$

4) $23 + 7x = 9$

5) $8x - 13 = 35$

6) $19 + 3x = 4$

7) $-1 + 7x = -22$

8) $5 - 2x = -21$

9) $-17 = 5 - 11x$

10) $-8 + 9x = 46$

Level 2:

1) $3(x + 5) = 24$

2) $5(x - 2) = -25$

3) $13 = 2(x + 3)$

4) $7 + x/3 = 11$

5) $9 = x/5 - 4$

6) $-4(x - 7) = 16$

7) $12(x - 7) = 108$

8) $-8 = 2(x - 5)$

9) $11 - x/3 = 4$

10) $-15 = x/9 - 13$

Level 3:

1) $3 = (18 - x)/4$

2) $6 = 42/(x - 5)$

3) $24/(2x) = 4$

4) $8 = (1/2)(x + 3)$

5) $(5x)/4 - 3 = 7$

6) $-11 = (9 + x)/2$

7) $-3 = 18/(x - 2)$

8) $49/(4x) = -7$

9) $-9 = (1/3)(x - 15)$

10) $(-8x)/2 + 5 = -7$

Level 4:

1) $(3x + 4)/2 = 11$

2) $5(x + 2) + 7 = 37$

3) $(8x + 3)/3 - 5 = 4$

4) $9 = (7 - 3x)/4 + 6$

5) $11 = 55/(7x - 16)$

6) $(4x - 6)/3 = -10$

7) $-6(x + 5) - 3 = -57$

8) $(11x - 10)/7 + 23 = 31$

9) $-1 = (-3 + 6x)/3 - 12$

10) $7 = 98/(-4x + 2)$

Level 5:

1) $2(x^2 + 3) - 7 = 49$

2) $13 - 24/(x + 3) = 9$

3) $5(4x - 13)^3 = 135$

4) $81/(x-5)^2 + 17 = 26$

5) $7 + (x + 2)^2 = 71$

6) $6(x^2 - 27) + 15 = -51$

7) $73 - 42/(x - 12) = 67$

8) $13(9x - 16)^3 = 104$

9) $112/(x-6)^2 - 5 = 23$

10) $-15 + (x - 9)^2 = 34$

Teaching Guide

Sticky Note Cover Up Challenge

App: Cover Up

Overview:

Students practice the Cover Up Math Method with pencil and paper problems that mirror the problems encountered in the iPad app *Cover Up*.

Common Core State Standards:
- CCSS.Math.Content.6.EE.A.2
- CCSS.Math.Content.6.EE.B.5
- CCSS.Math.Content.6.EE.B.6
- CCSS.Math.Content.6.EE.B.7
- CCSS.Math.Content.7.EE.B.3
- CCSS.Math.Content.8.EE.C.7
- CCSS.Math.Content.HSA.SSE.A.1
- CCSS.Math.Content.HSA.REI.A.1

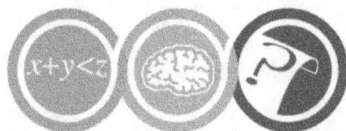

Encyclopedia of Algebraic Thinking:
- Algebraic Relations: Equations Involving Negatives
- Algebraic Relations: Flexible Use of Solution Strategies
- Algebraic Relations: One and two step equations
- Algebraic Relations: Student Intuition and Informal Procedures
- Variables and Expressions: Letter Used as a Specific Unknown

Description:

This activity is designed as a follow up to the app *Cover Up*. The purpose is to help students transfer the skills they learned while using the app to a pencil and paper activity so that they will then be more likely to use the method effectively when working on their own problems. Key:

Level 1 - **1**: 5; **2**: 2; **3**: 4; **4**: -2; **5**: 6; **6**: -5; **7**: -3; **8**: 13; **9**: 2; **10**: 6
Level 2 - **1**: 3; **2**: -3; **3**: 3.5; **4**: 12; **5**: 65; **6**: 3; **7**: 16; **8**: 1; **9**: 21; **10**: -18
Level 3 - **1**: 6; **2**: 12; **3**: 3; **4**: 13; **5**: 8; **6**: -31; **7**: -4; **8**: -7/4; **9**: -12; **10**: 3
Level 4 - **1**: 6; **2**: 4; **3**: 3; **4**: -5/3; **5**: 3; **6**: -6; **7**: 4; **8**: 6; **9**: 6; **10**: -3
Level 5 - **1**: 5; **2**: 3; **3**: 4; **4**: 8; **5**: 6; **6**: 4; **7**: 19; **8**: 2; **9**: 8; **10**: 16

Extensions:
- Comparison of Methods: Cover Up VS Algorithmic
- The Mental Math Cover Up Challenge

Comparison of Methods:
Cover Up VS Algorithmic

For use with the iPad app *Cover Up*

Name: Date: Period:

In this assignment you will compare two methods of solving equations: the cover up method and the algorithmic method.

Using the Two Methods

Solve the following problems using the **algorithmic** method:

1) $5 - 2x = -21$ 2) $-11 = (9 + x)/2$

Now solve these problems using the **cover up** method:

3) $19 + 3x = 4$ 4) $-3 = 18/(x + 2)$

Reflecting on the Two Methods

What is different about these two methods?

Which of the two methods do you prefer using? Explain your choice.

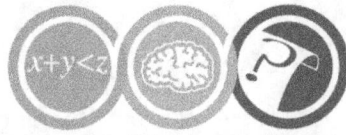

Teaching Guide

Comparison of Methods: Cover Up vs. Algorithmic

App: Cover Up

Overview:

In this activity students solve a few equations using the cover up method and a few using the algorithmic method, and then reflect on the differences between the two methods.

Common Core State Standards:

- CCSS.Math.Content.6.EE.A.2
- CCSS.Math.Content.6.EE.B.5
- CCSS.Math.Content.6.EE.B.6
- CCSS.Math.Content.6.EE.B.7

- CCSS.Math.Content.7.EE.B.3
- CCSS.Math.Content.8.EE.C.7
- CCSS.Math.Content.HSA.SSE.A.1
- CCSS.Math.Content.HSA.REI.A.1

Encyclopedia of Algebraic Thinking:
- Algebraic Relations: Equations Involving Negatives
- Algebraic Relations: Flexible Use of Solution Strategies
- Algebraic Relations: One and two step equations
- Algebraic Relations: Student Intuition and Informal Procedures
- Variables and Expressions: Letter Used as a Specific Unknown

Description:

The purpose of this activity is to give students the opportunity to reflect on two different methods for solving equations. When we refer to the "algorithmic method" we are referring to the step by step process of solving an equation by executing identical operations on both sides of the equal sign: (from problem 1 in this activity) subtract 5 from both sides to isolate the term with x in it, then divide both sides by -2 to get x = 13. When we refer to the "cover up method," we are referring to the method introduced in the activity Sticky Note Cover Up Math that goes like this: (from problem 3 in this activity) in order to get 4 from 19 I have to subtract 15, then in order to get 15 from 3 I have to multiply by 5. Both methods have their merits, and the purpose of this activity is to point out those merits and give students multiple approaches to solving equations.

Extensions:

- The Mental Math Cover Up Challenge

The Mental Math Cover Up Challenge

For use with the iPad app *Cover Up*

Name: Date: Period:

In this activity you will test your mental math skills by seeing how quickly you can use the cover up math method to solve equations without writing anything down!

Use the cover up math method **in your head** to solve the following problems:

Level 1:

1) $5x + 7 = 32$ 2) $12 - 3x = 6$

3) $9 = 25 - 4x$ 4) $23 + 7x = 9$

5) $8x - 13 = 35$ 6) $19 + 3x = 4$

7) $-1 + 7x = -22$ 8) $5 - 2x = -21$

9) $-17 = 5 - 11x$ 10) $-8 + 9x = 46$

Level 2:

1) $3(x + 5) = 24$ 2) $5(x - 2) = -25$

3) $13 = 2(x + 3)$ 4) $7 + x/3 = 11$

5) $9 = x/5 - 4$ 6) $-4(x - 7) = 16$

7) $12(x - 7) = 108$ 8) $-8 = 2(x - 5)$

9) $11 - x/3 = 4$ 10) $-15 = x/9 - 13$

Level 3:

1) $3 = (18 - x)/4$

2) $6 = 42/(x - 5)$

3) $24/(2x) = 4$

4) $8 = (1/2)(x + 3)$

5) $(5x)/4 - 3 = 7$

6) $-11 = (9 + x)/2$

7) $-3 = 18/(x - 2)$

8) $49/(4x) = -7$

9) $-9 = (1/3)(x - 15)$

10) $(-8x)/2 + 5 = -7$

Level 4:

1) $(3x + 4)/2 = 11$

2) $5(x + 2) + 7 = 37$

3) $(8x + 3)/3 - 5 = 4$

4) $9 = (7 - 3x)/4 + 6$

5) $11 = 55/(7x - 16)$

6) $(4x - 6)/3 = -10$

7) $-6(x + 5) - 3 = -57$

8) $(11x - 10)/7 + 23 = 31$

9) $-1 = (-3 + 6x)/3 - 12$

10) $7 = 98/(-4x + 2)$

Level 5:

1) $2(x^2 + 3) - 7 = 49$

2) $13 - 24/(x + 3) = 9$

3) $5(4x - 13)^3 = 135$

4) $81/(x-5)^2 + 17 = 26$

5) $7 + (x + 2)^2 = 71$

6) $6(x^2 - 27) + 15 = -51$

7) $73 - 42/(x - 12) = 67$

8) $13(9x - 16)^3 = 104$

9) $112/(x-6)^2 - 5 = 23$

10) $-15 + (x - 9)^2 = 34$

Center for Algebraic Thinking

Teaching Guide

The Mental Math Cover Up Challenge

App: Cover Up

Overview:

In this activity students are challenged to see if they can use the cover up math method to solve equations in their head.

Common Core State Standards:

- CCSS.Math.Content.6.EE.A.2
- CCSS.Math.Content.6.EE.B.5
- CCSS.Math.Content.6.EE.B.6
- CCSS.Math.Content.6.EE.B.7

- CCSS.Math.Content.7.EE.B.3
- CCSS.Math.Content.8.EE.C.7
- CCSS.Math.Content.HSA.SSE.A.1
- CCSS.Math.Content.HSA.REI.A.1

Encyclopedia of Algebraic Thinking:

- Algebraic Relations: Equations Involving Negatives
- Algebraic Relations: Flexible Use of Solution Strategies
- Algebraic Relations: One and two step equations
- Algebraic Relations: Student Intuition and Informal Procedures
- Variables and Expressions: Letter Used as a Specific Unknown

Description:

Once students have practiced using the cover up math method with pencil, paper, and sticky notes, then they might be up for the challenge of solving equations in their head using the same method. This activity gives them the opportunity to take on that challenge. Note that the equations in this activity are the same as those in the activity Sticky Note Cover Up Challenge.

Extensions:

If students are enjoying the challenge of using the cover up math method to solve equations in their head, then you could add some friendly competition into the mix by holding a class tournament to see who is the fastest at solving equations using mental math. While two students at a time go head-to-head, the rest of the class could also participate by recording their answers on a sheet of paper (whereas the competitors would share their answers vocally - maybe by buzzing in first).

81

Diamond Factor

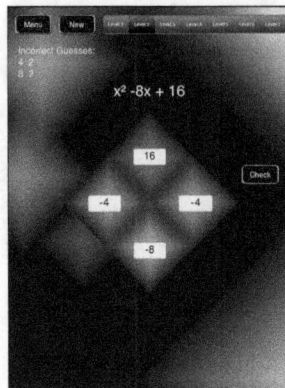

Diamond Factor helps you to learn how to factor trinomials of various difficulties through the use of a diamond. Players can practice levels which range from trinomials with small positive coefficients to larger, negative or decimal coefficients. Players can also factor trinomials in game mode which keeps track of accuracy and time. This is a standard approach to factoring used in schools that pushes students to consider the relationship between the factored and unfactored forms of a trinomial, particularly in regard to how multiplication and addition create the terms.

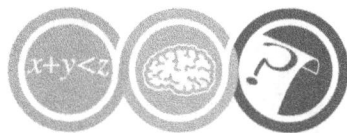

Center for Algebraic Thinking

Deriving the Diamond Method: Part 1

For use with the iPad app *Diamond Factor*

Name: Date: Period:

In this assignment you will learn about the Diamond Method of factoring quadratic equations. After you learn this method you will get to practice it using the iPad app Diamond Factor.

1. Use the Distributive Property or the FOIL method to show that $(x + 3)(x + 2) = x^2 + 5x + 6$:

    Note: You may not need all of the lines provided

    $(x + 3)(x + 2)$ = _____

                           = _____

                           = _____

                           = $x^2 + 5x + 6$

2. The goal of **factoring** is to take an expression such as $x^2 + 5x + 6$ and "reverse distribute" it to get to $(x + 3)(x + 2)$. Your next task is to develop a method that will allow you to factor expressions in that way. The key is finding the connection between the bold terms on the left side of the equation and the bold terms on the right side of the equation:

    $$(x + \mathbf{3})(x + \mathbf{2}) = x^2 + \mathbf{5}x + \mathbf{6}$$

    What connection do you see between the two bold numbers on the left and the two bold numbers on the right of the equation above?

3. Now it's time to test the connection that you developed. See if it holds true for the following equations. If it doesn't, keep modifying the connection until it works for ALL of the equations.

    a. $(x + 4)(x + 1) = x^2 + 5x + 4$ **b.** $(x + 2)(x + -3) = x^2 + -1x + -6$

    c. $(x - 6)(x + 4) = x^2 - 2x - 24$ **d.** $(x - 7)(x - 2) = x^2 - 9 + 14$

Teaching Guide

Deriving the Diamond Method: Part 1

App: Diamond Factor

Overview:

In this activity students set the stage for learning the Diamond Method as they discover for themselves the connection between the factored form and foiled form of a quadratic equation.

Common Core State Standards:

- CCSS.Math.Content.HSA.SSE.B.3
- CCSS.Math.Content.HSF.IF.C.8

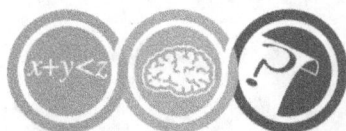

Encyclopedia of Algebraic Thinking:

- Algebraic Relations: Student Intuition and Informal Procedures
- Patterns and Functions
- Variables and Expressions: Letter Used as a Specific Unknown

Description:

The purpose of this assignment is both to remind students about FOILing (i.e. distributing) a factored equation and to give them the chance to discover for themselves why the diamond method works before they learn the mechanics of that method. This approach is potentially more worthwhile than simply *telling* them what the diamond method is because if they know where the method comes from then they will see that tools like these have a logical origin that they themselves can often discover. Students will also be more likely to remember the method after having seen where it comes from.

Extensions:

- Discovering the Diamond Method: Part 2

Center for Algebraic Thinking

Deriving the Diamond Method: Part 2

For use with the iPad app *Diamond Factor*

Name: Date: Period:

In Part 2 of this assignment you will take the connection that you developed in Part 1 and solidify that connection using the Diamond Method.

In Part 1 you should have found that to factor an equation such as $x^2 + 5x + 6$, you need to find two numbers that add to 5 and multiply to 6. The two numbers that satisfy those conditions are 2 and 3, so: $x^2 + 5x + 6 = (x + 2)(x + 3)$.

We can *generalize* that relationship in the following way: If we have an equation of the form

$$x^2 + bx + c$$

Then in order to factor that expression we need to find two numbers that add together to get b and multiply together to get c. Let's call the two numbers that meet those conditions f and g. What you found out in Part 1 of this assignment is that

IF

$f + g = b$ and $f*g = c$

THEN

$x^2 + bx + c = (x + f)(x + g)$

To help us remember that relationship and use it to solve problems, we have the Diamond Method, which works like this:

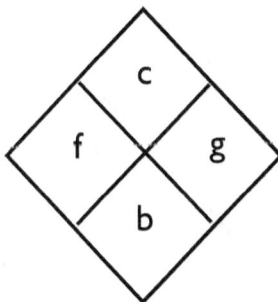

The two numbers in the left and right sections of the diamond have to add together to get the number in the bottom section of the diamond and multiply together to get the number in the top section of the diamond. Check out the next page for a few examples.

85

EXAMPLES:

Ex 1.
$$x^2 + 5x + 6$$

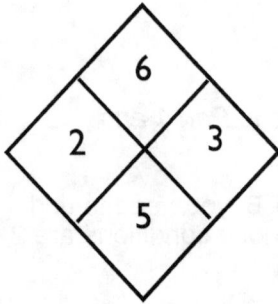

$$(x + 2)(x + 3)$$

Ex 2.
$$x^2 + 5x + 4$$

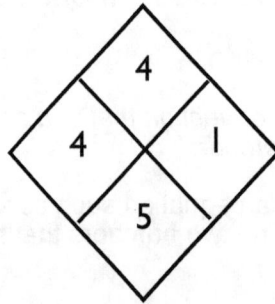

$$(x + 4)(x + 1)$$

Ex 3.
$$x^2 - 2x - 24$$

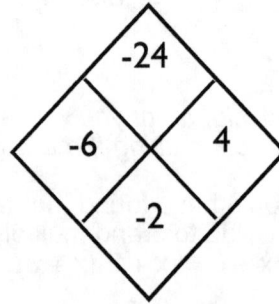

$$(x - 6)(x + 4)$$

Now it's time for you to get some practice using this method. See if you can fill in the missing sections of these diamonds, and then write in the factors below the diamond:

1.
$$x^2 + 11x + 28$$

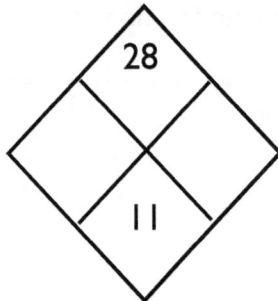

(x   )(x   )

2.
$$x^2 - 2x - 15$$

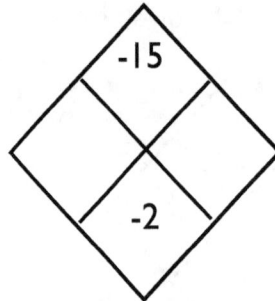

(x   )(x   )

3.
$$x^2 + 3x - 54$$

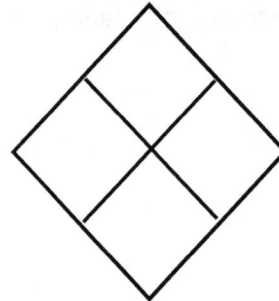

(x   )(x   )

Now you get to practice this new skill using the app *Diamond Factor*. Your goal is to get all the way to Level 5. Open it up and see how you do!

Teaching Guide

Deriving the Diamond Method: Part 2

App: Diamond Factor

Overview:

In this activity students build on the foundation set in Part 1 as they learn the mechanics of the Diamond Method and then practice using that method.

Common Core State Standards:

- CCSS.Math.Content.HSA.SSE.B.3
- CCSS.Math.Content.HSF.IF.C.8

Encyclopedia of Algebraic Thinking:

- Algebraic Relations: Student Intuition and Informal Procedures
- Patterns and Functions
- Variables and Expressions: Letter Used as a Specific Unknown

Description:

In this follow-up to Part 1, students are walked through the process of *generalizing* the connection that they (hopefully) made between the factored form and distributed form of a quadratic equation. The mechanics of the Diamond Method are then introduced, and students get a chance to practice using that method with pencil and paper before taking on the challenges in the app *Diamond Factor*.

Extensions:

After learning this method students can work through the "Play" section of the app *Diamond Factor* and see how many levels they can pass through in a given amount of time.

87

Function Mystery Machine

A simple, fun way to practice algebraic functions. Choose a level or go head-to-head with a friend as you try to guess the mystery function. The game supports algebra functions ranging from simple "x + 5" equations to ones such as quadratics and "x + a" for two-players. Developing hypotheses and testing those hypotheses helps students understand the role of variables within a function and the influence of order of operations upon those variables.

Center for Algebraic Thinking

Discovering the Mystery Function

For use with the iPad app *Function Mystery Machine*

For this activity you will use the iPad app *Function Mystery Machine* to test your knowledge of linear functions and see if you can outsmart your classmates in the two-player challenge. Complete Phase 1 below before moving to Phase 2.

Phase 1: One-Player Practice

Steps:

1. Work your way up through Level 4 of the app, to the start of Level 5.
2. Show the iPad to your instructor to prove that you are ready to move on to Phase 2.

Phase 2: Two-Player Challenge

Steps:

1. From the Menu, Player 1 writes in a linear equation and taps 'Start Two-Player'.
2. Player 2 has 5 attempts to write in the correct equation. Player 2 gains a point from writing in the correct equation, but loses a point if unable to write the correct equation within 5 attempts.
3. Player 2 writes in a linear equation.
4. Player 1 has 5 attempts to write in the correct equation. Player 1 gains a point from writing in the correct equation, but loses a point if unable to write the correct equation within 5 attempts.
5. Continue alternating roles until a player reaches 5 points - this player is the winner!
6. Once a player reaches 5 points, show the iPad to the Instructor as a check for participation.

89

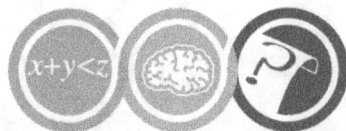

Teaching Guide

Discovering the Mystery Function

App: Function Mystery Machine

Overview:

In this activity students use the iPad app *Function Mystery Machine* to first practice individually the process of identifying a linear equation based on a table of points, and then use the two-player mode to challenge a classmate.

Common Core State Standards:

- CCSS.Math.Content.8.EE.C.7
- CCSS.Math.Content.8.F.A.1
- CCSS.Math.Content.8.F.B.4
- CCSS.Math.Content.HSA.CED.A.1

- CCSS.Math.Content.HSA.REI.B.3
- CCSS.Math.Content.HSF.BF.A.1
- CCSS.Math.Content.HSF.LE.A.2

Encyclopedia of Algebraic Thinking:

- Analysis of Change: Understanding Slope
- Patterns and Functions: Function Machine
- Patterns and Functions: Linear Function

Description:

In this two-phase activity, students first show that they have mastered individually the skill of defining a linear function based on a table of points, then move on to a two-player challenge in which students write equations for each other to discover. The strength of this app is that the two-player mode is built into the application, so students need only to follow the prompts provided by the app. The steps listed in this activity are simply an additional reference.

Extensions:

If both students participating in the two-player challenge have completely mastered linear equations, then they can start incorporating quadratic and other complex functions. One-Player Levels 5 and above incorporate more complex functions if students need individual practice first.

Hop the Number Line

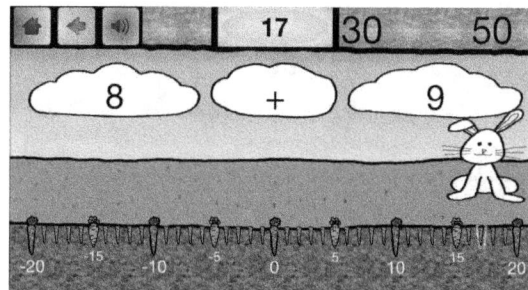

Research shows that students struggle with negative numbers. In this app, students drag the Bunny across the number line of carrots to answer addition and subtraction problems in this fun race against the clock math game. Students struggle with integers and understanding negative numbers, in particular. This app helps students practice combinations of integers.

Levels 1-4: Two term integer equations

Levels 4-8: Three term integer equations

Levels 9-11: Three term decimal and fraction equations

91

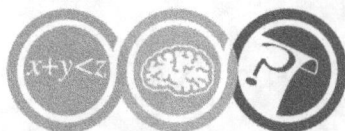

Center for Algebraic Thinking
An Introduction

For use with the iPad app *Hop the Number Line*

Name: Date: Period:

In this activity you will practice a method used to solve addition and subtraction problems. After practicing this method on paper, you'll use it in the app Hop the Number Line.

Example 1: -4 + 9

Step 1: Add a point at the location of the first number, and label it with an 'A'. This is your starting point.

-4 + 9

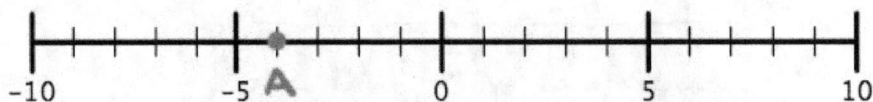

Step 2: Based on the operation and the number that follows the first, draw an arrow that shows the number of steps and the direction you must move the point. Label this second location with a 'B'. The location of 'B' is your final answer!

-4 + 9

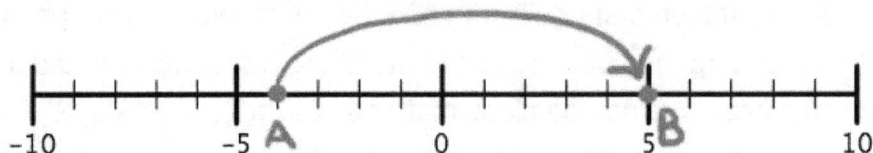

Final Answer: 5

Example 2: 3 - 5 + 8

Steps 1 and 2: Same as above.

3 - 5 + 8

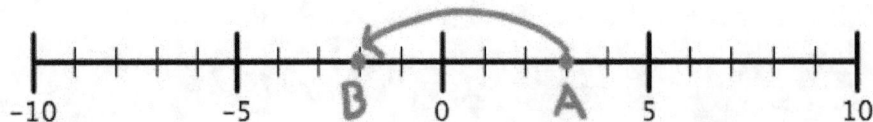

Step 3: Since we have one more operation, draw another arrow that shows where you need to move the dot to the final answer. Label this final location with a 'C'. The location of the point 'C' is your final answer!

$$3 - 5 + 8$$

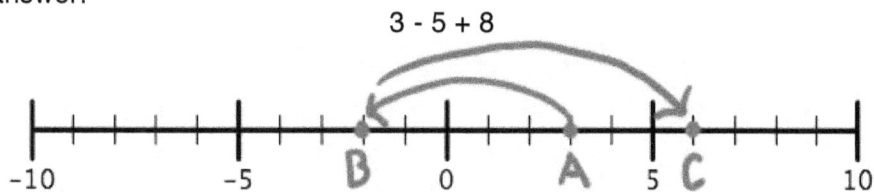

Final Answer: 6

Practice

Now it's your turn to give this method a try. Using the examples above as your model, solve the following problems in the same way. Make sure that for each problem you first draw it out on the number line, and then write in your final answer below it.

BeginnerProblems

1.   8 - 6

Final Answer: _____

2.   7 - 13

Final Answer: _____

3.   -5 + 11

Final Answer: _____

4.   3 + -7

Final Answer: _____

93

Intermediate Problems

1. 5 - 7 - 4

Final Answer: _____

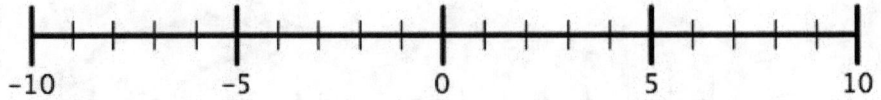

2. 3 + -10 + 2

Final Answer: _____

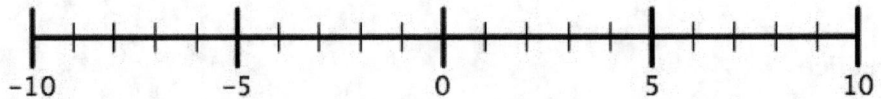

3. -6 + 4 - -9

Final Answer: _____

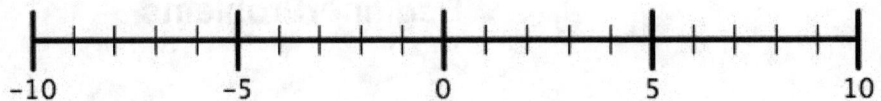

4. -7 - -11 - 3

Final Answer: _____

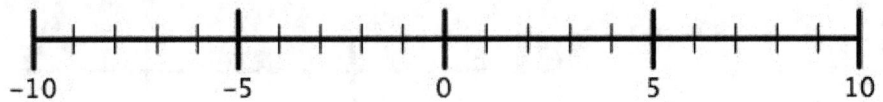

Advanced Problems

5. 1.1 - 1.5 + 0.2

Final Answer: _____

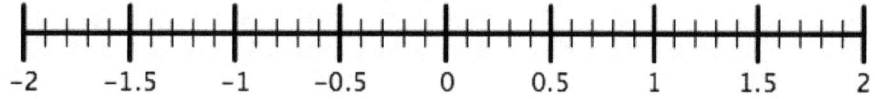

6. 1.1 - 1.5 + 0.2

Final Answer: _____

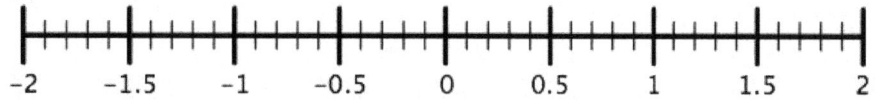

7. 1.1 - 1.5 + 0.2

Final Answer: _____

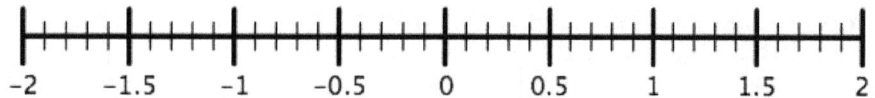

8. 1.1 - 1.5 + 0.2

Final Answer: _____

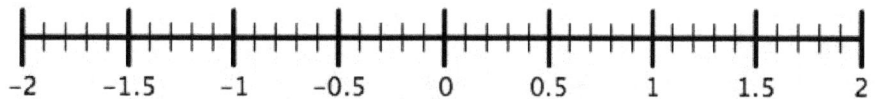

95

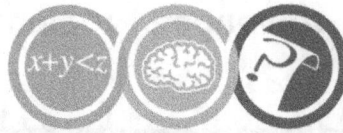

Center for Algebraic Thinking

Teaching Guide

Hop the Number Line: An Introduction

App: Hop the Number Line

Overview:

In this activity students practice solving addition and subtraction problems on paper using a number line. After practicing this method on paper students will be prepared to use it with the app *Hop the Number Line*.

Common Core State Standards:
- CCSS.Math.Content.6.NS.B.3
- CCSS.Math.Content.6.NS.C.6
- CCSS.Math.Content.7.NS.A.1

Encyclopedia of Algebraic Thinking:
- Algebraic Relations: Flexible Use of Solution Strategies
- Algebraic Relations: Negative Numbers

Description:
Using the number line to solve addition and subtraction problems gives students a useful and concrete visual representation of the operations that they are performing. It removes some of the abstractness that can plague higher levels of mathematics. This activity helps students make and solidify the link between basic operations and how those operations are represented on the number line. Representing problems in this way can be particularly useful as students encounter negative numbers and the addition and subtraction of negative numbers. The app *Hop the Number Line* provides great practice as well, but this introductory activity will help students who aren't as familiar with using the number line to solve these types of problems. The different levels of practice problems in this activity (Beginner, Intermediate, and Advanced) match the levels that are found in the app, and provide enough range of difficulty to be beneficial for many students.

Extensions:
Once students are comfortable with this method, they can simply start using the app and moving through the levels. For a whole class activity, take a look at Hop the Number Line: Student Hop Challenge. For additional methods and challenges, see Hop the Number Line: Three Approaches.

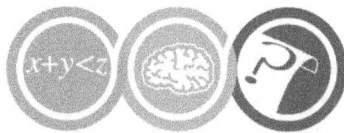

Center for Algebraic Thinking

Student Hop Challenge

For use with the iPad app *Hop the Number Line*

In this activity students play the role of the bunny in the app Hop the Number Line *in front of the class, literally hopping as they work through an addition or subtraction problem with the help of classmates when necessary.*

Setup

Draw a large number line across the board in the front of the room. Have a set of addition and subtraction problems on hand (with two, three, or four steps) with a range of difficulties. You could also make the problems up on the spot if you're comfortable with that.

Activity

Request a volunteer from the class. That student goes to the front of the room. Write a problem on the board appropriate for that student. The student then follows the method outlined in Hop the Number Line: An Introduction, but using his or her physical body to demonstrate moving up or down the number line. To add a physical challenge, you could encourage students to see if they can make it from one point to the next in a single hop. You could encourage class participation by telling the rest of the class that they can clap if they agree with the hop, or silently raise their hand or point in the correct direction if they think the hop was made in error. By writing the number line on the board (as opposed to doing something with tape on the floor), the entire class can more easily see, and the active student simply has to stand in front of the appropriate mark on the board.

As the instructor, you can choose how many problems to work through in this way, as well as the degree of difficulty of each problem.

Teaching Guide

Student Hop Challenge

App: Hop the Number Line

Overview:

In the activity students physically play the role of the bunny in the app *Hop the Number Line*, solving an addition or subtraction problem in front of the class by hopping from one point to the next along a number line drawn large enough for the whole class to see.

Common Core State Standards:

- CCSS.Math.Content.6.NS.B.3
- CCSS.Math.Content.6.NS.C.6
- CCSS.Math.Content.7.NS.A.1

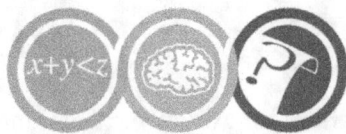

Encyclopedia of Algebraic Thinking:

- Algebraic Relations: Flexible Use of Solution Strategies
- Algebraic Relations: Negative Numbers

Description:

This activity is fully described on the previous page (that document is meant for instructors only, not to be handed out to students). It engages students in physical activity while also encouraging whole-class participation.

Extensions:

- Hop the Number Line: An Introduction

- Hop the Number Line: Three Approaches

Center for Algebraic Thinking

Three Approaches

For use with the iPad app *Hop the Number Line*

Name: Date: Period:

In this activity you will test out three different approaches to the game Hop the Number Line, *then pick the one you like best and see how high of a score you can get.*

Part 1: Testing the Three Approaches

From the home screen of the app, tap 'PLAY,' then 'Practice,' then select a level that seems neither too easy not to difficult. Then play that level using each of the approaches below, and answer the questions about how it goes.

Approach 1: Fly Bunny, Fly

Begin by moving the bunny to location on the number line that is shown in the first cloud. Think of that as your starting point - that's where the bunny starts her (or his) journey. If the number in the first cloud is off of the screen (too big or too small) then just move the bunny as far as you can to that side of the screen and imagine that she is all the way out at the correct number.

Then look at the next two clouds, and move the bunny the correct number of steps in the appropriate direction. If there are no more clouds after that, you're done and you can tap the green box with the number in it to see if you correct! If you have more clouds then move the bunny one more time, then check your answer.

Follow the two steps above until you've correctly completed 5 problems. Then answer the questions below:

1) What did you think of using this method?

2) Did you have any difficulties with this method? If so, explain.

99

Approach 2: Write It Down

Make sure you have some scratch paper on hand. Use that scratch paper to write out the problem in the clouds by hand, and then solve it. Once you're happy with your answer, move the bunny the appropriate spot and tap the green box to see if you're correct! Continue using that process until you've correctly completed 5 problems. Then answer the questions below:

1) What did you think of using this method?

2) Did you have any difficulties with this method? If so, explain.

Approach 3: Mental Math

Without moving the bunny or writing anything down, complete the cloud problem *in you head*, then move the bunny to the correct spot and check your answer. Continue using that process until you've correctly completed 5 problems. Then answer the questions below:

1) What did you think of using this method?

2) Did you have any difficulties with this method? If so, explain.

BONUS: Do you have any ideas for a different way to solve the problems in this app? If so, share your idea below:

Part 2: Game On

From the methods you tried out on the previous pages, pick your favorite one, then play a scored game (from the home screen tap 'PLAY' then 'Scored') at a level that challenges you but isn't so hard that you can't score any points. When the timer runs out, record the level and the number of points you scored below. Then try a new level and play again (or play the same level and try to beat your previous score!). As you get better, see if you can advance to harder and harder levels.

Level: _____   Score: _____

Level: _____   Score: _____

Level: _____   Score: _____

Level: _____   Score: _____

Level: _____   Score: _____

Level: _____   Score: _____

Level: _____   Score: _____

Level: _____   Score: _____

Level: _____   Score: _____

Level: _____   Score: _____

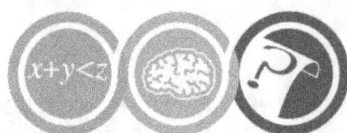

Center for Algebraic Thinking

Teaching Guide

Hop the Number Line: Three Approaches

App: *Hop the Number Line*

Overview:

In this activity students learn about three different approaches to solving the addition and subtraction problems presented in the app *Hop the Number Line*, and then pick the approach they like best and play the game repeatedly with that method, advancing to harder levels as they improve their skills.

Common Core State Standards:
- CCSS.Math.Content.6.NS.B.3
- CCSS.Math.Content.6.NS.C.6
- CCSS.Math.Content.7.NS.A.1

Encyclopedia of Algebraic Thinking:
- Algebraic Relations: Flexible Use of Solution Strategies
- Algebraic Relations: Negative Numbers

Description:

This activity reinforces the idea that any given problem can be solved in a variety of ways. In Part 1, students try out three different methods and write down what they think of each method, including whether or not they had any difficulties using it. In Part 2, students select the method they like the best and use that method to play scored games with the goal of improving their score with each new round and advancing to more difficult levels. The variety of the methods, in addition to the varied difficulty of the levels, allow for students of all abilities and problem solving styles to learn something new.

Extensions:

Because of the increased difficulty with each new level, students should be encouraged to try harder levels (as well as harder methods - such as the mental math approach) once they have mastered a given level or method.

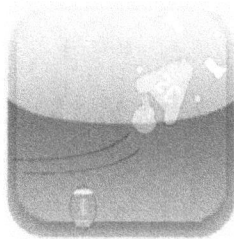

Inequality Kickoff

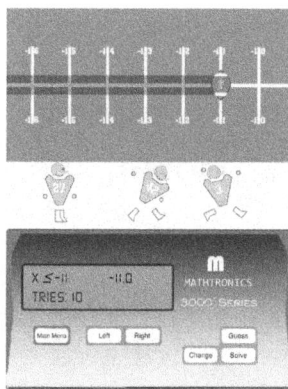

In **Inequality Kickoff** it is your job to set up the football on the field for the perfect kick! Set up the ball according to the given inequality and let the pigskin fly! Guess right and you'll have that ball sailing all day! Get it wrong and your kicker will be the one taking the heat. 12 levels of football themed inequality solving that help students understand the meaning by using inequalities on a number line. We know from research that students have significant difficulty understanding which direction to shade lines and the meaning of that shaded line. They struggle to understand how values work in an inequality. This program helps students practice thinking about those concepts.

103

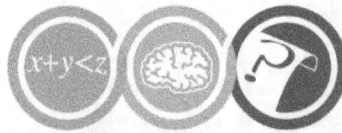

Center for Algebraic Thinking

A New Football Field
With a New Coordinate System

For use with the iPad app *Inequality Kickoff*

Name: Date: Period:

In this activity you will investigate what happens when you change the coordinate system used to describe a football field.

On the next page you will see two football fields. The first shows what a football field actually looks like. The second shows an alternate version of a football field with a different method of numbering the yard lines. Take a look at both fields, then answer the questions below.

Questions

1. Assuming the game is still played the same (one team moving across the field left to right, the other team moving right to left), what would be different about the way the game is *described*?

Let's use x to describe the location of the football in the new system. For the following questions, assume that the team we're rooting for is moving from left to right across the field.

2. If we're on our own 20 yard-line (as described on the old field), where would we be on the new field (make sure to use x in your response)?

3. If we're at midfield, what would our location be in the new system of describing football fields? (Again, make sure to use x in your response)

4. If we're on the opponent's 5 yard-line (as described on the old field), where would we be on the new field?

5.   a. In the old system, what does it mean to be in the 'red zone'?

       b. How would you describe the red zone in the new system (using x)?

The questions on the previous page talked about locations on the field. Now let's talk about plays. For all of the questions below, assume that our team is moving left to right across the field. The first descriptions use the old system for describing a football field, and your task is to describe that situation in the new system.

6. Let's say that we start at our own 20, complete a pass, and the receiver gets tackled at the 50 yard line. How could you describe the ground that was covered in our new system, using an inequality?

7. Now we're at the 50, and we run the ball all the way to our opponent's 15 yard line. How could you describe the ground covered in our new system, using an inequality?

8. Have you seen the movie Forest Gump? There's a scene in that movie where Forest returns a kickoff for a touchdown, but then just keeps on running. Think about how you would describe that situation. Let's say he catches the ball at his own 5 yard-line, returns it for a touchdown, then just keeps running, right out of the stadium and continuing in the same direction. How would you describe the ground he covers in our new system, using an inequality?

9. Now it's your turn to make up your own play. Describe a play using the old system (the one actually used in football games). Then write down how you would describe that same play in the new system (using x).

105

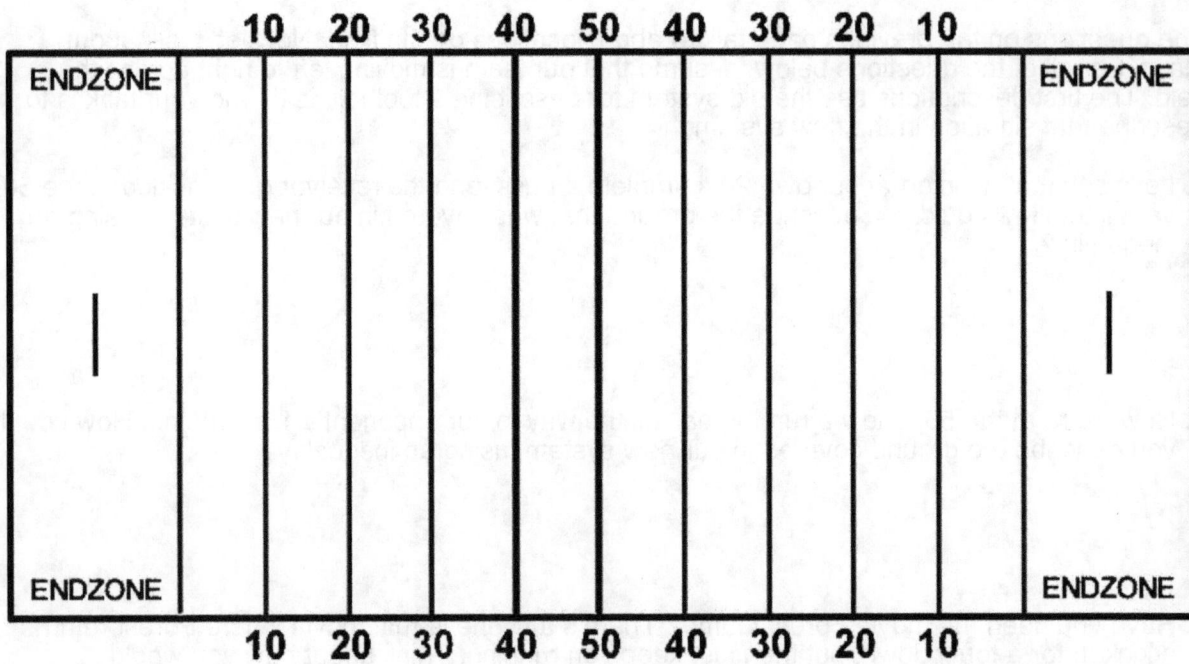

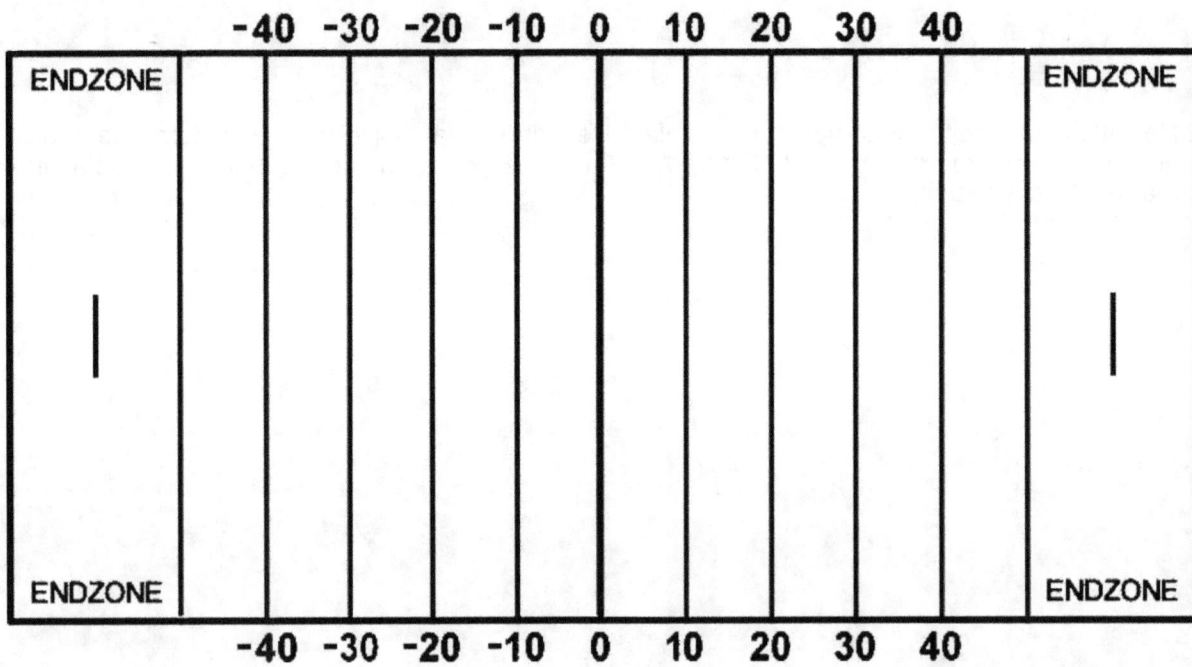

Center for Algebraic Thinking

Teaching Guide

A New Football Field

App: Inequality Kickoff

Overview:
The purpose of this activity is to introduce students to the idea of applying a different coordinate system to a football field than the one that is typically used. This can be a stand-alone activity, but it also works well as a primer for the app Inequality Kickoff (which employs the new coordinate system that this activity introduces). This particular activity does not use the app at all, but the extension activity does (see below).

Common Core State Standards:
- CCSS.Math.Content.6.NS.C.5
- CCSS.Math.Content.6.NS.C.6
- CCSS.Math.Content.6.EE.B.8
- CCSS.Math.Content.7.EE.B.4
- CCSS.Math.Content.HSA.CED.A.1

Encyclopedia of Algebraic Thinking:
- Algebraic Relations: Inequalities
- Modeling: Translating Word Problems into Equations

Description:
Since not everyone will be familiar with the game of football, it might be best to do this activity as an entire class (or in small groups so long as at least one person in the group is familiar with football terminology). The main thing to emphasize here is that we as mathematicians get to decide what coordinate system we apply to different situations. Over time we might come to a general consensus about which coordinate system we want to use, but there's nothing stopping us from trying something different if we think it might work better for our purposes. In the case of this activity (and the app Inequality Kickoff), we look at what happens when you apply the standard number line to a football field, with zero at midfield, negative numbers to the left, and positive numbers to the right. Through a series of questions students investigate what it would be like if football fields were actually marked in that way, and how they would describe the game using that new system.

Extensions:
- Inequality Kickoff Progress Monitor
- Make up your own game using the new coordinate system for a football field

107

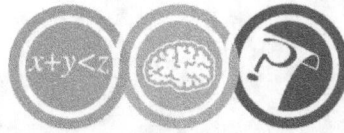

Center for Algebraic Thinking

Inequality Kickoff Progress Monitor

For use with the iPad app *Inequality Kickoff*

Name: _____ Date: _____ Period: _____

Use the space below to record your correct solutions as you play the game Inequality Kickoff. As you play the game, make sure to tap 'Guess' (NOT 'Solve') so that you can still see the problem, your solution, and whether or not it is correct. Once you get the "CORRECT!" message you can record your solution below.

Level: _____ Problem: _____

Solution:

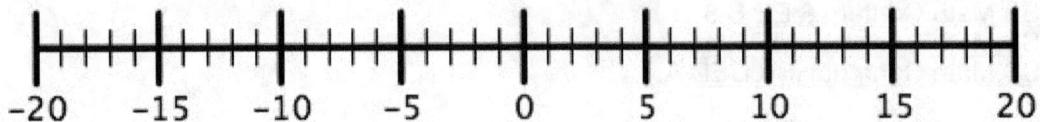

Level: _____ Problem: _____

Solution:

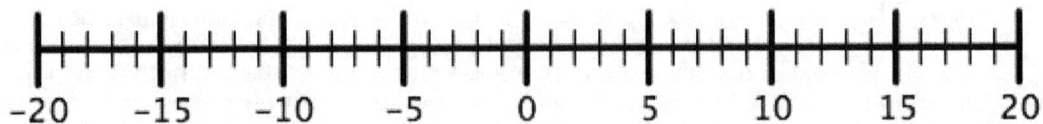

Level: _____ Problem: _____

Solution:

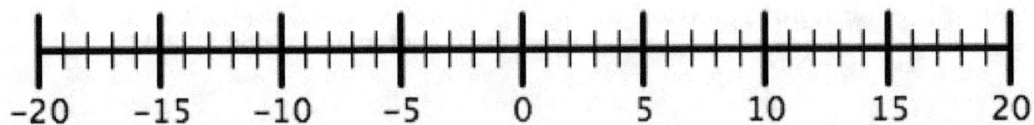

Level: _____   Problem: _____

Solution:

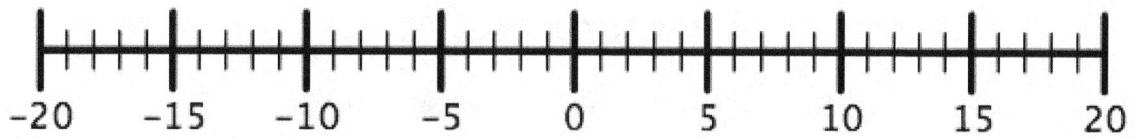

Level: _____   Problem: _____

Solution:

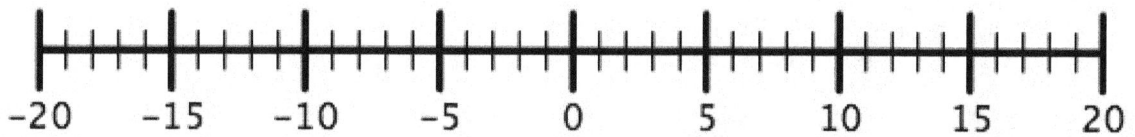

Level: _____   Problem: _____

Solution:

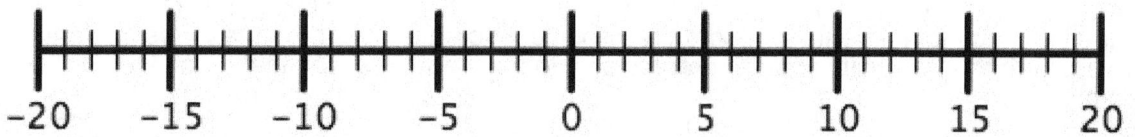

Level: _____   Problem: _____

Solution:

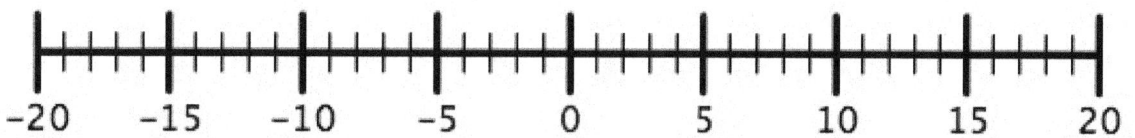

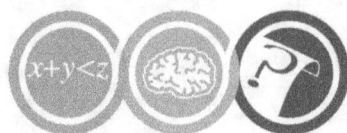

Teaching Guide

Inequality Kickoff Progress Monitor

App: Inequality Kickoff

Overview:

This paper resources provides students with a space to record their solution to each problem they face in the app *Inequality Kickoff*.

Common Core State Standards:
- CCSS.Math.Content.6.NS.C.5
- CCSS.Math.Content.6.NS.C.6
- CCSS.Math.Content.6.EE.B.8
- CCSS.Math.Content.7.EE.B.4
- CCSS.Math.Content.HSA.CED.A.1

Encyclopedia of Algebraic Thinking:
- Algebraic Relations: Inequalities
- Modeling: Translating Word Problems into Equations

Description:

In order to reinforce what is learned while using the app Inequality Kickoff, this resource provides a space for students to record their solutions to each problem. It requires students to write out both the inequality and the graphic solution by hand in a way that leads them to becoming more accustomed to the format. You as the instructor can decide which levels students should play and how they should progress from one level to the next. In addition to the reinforcement for the student, this resources also serves as an accountability tool - they can turn it in at the end of the period so that you can see how much work was done and how correctly the problems and solutions were transferred to paper.

Extensions:

Since each level progresses in difficulty, you can encourage students to continue to the higher levels if they find the lower levels too easy. If the student is familiar with the game of football, you could have them try A New Football Field.

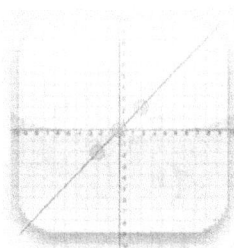

Linear Model

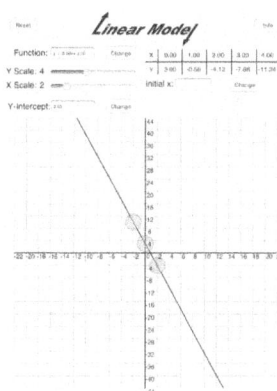

Students that learn how multiple representations of mathematical ideas fit together have more comprehensive understanding of a concept. Linear Model has an adjustable line, a function input and y-intercept input, and an x-y table. Any adjustment of the line, equation, or table automatically changes the other two representations so students can see how they are connected. This linear grapher has several options and features: an adjustable line, a function input and y-intercept input, and an x-y table. The adjustable line has three points along it. The point in between allows the user to move the line when touched and dragged, changing the y-intercept and the base value of the function. The other two points allow the user to rotate the line, also changing the y-intercept and changing the multiplier of the function. Aside from changing the graph via the line itself, the user also has the option of changing the line by entering values in the y-intercept field and in the function field. If the user enters a function that is not in y=mx+b form or in ax+by=c form, the app will not utilize and will give the user a warning that the app cannot use the function.

111

Drag and Drop Linear Equations

For use with the iPad app *Linear Model*

Name: Date: Period:

1. Use the central blue dot on the graph to drag the line around. What happens to the equation as you drag that line around?

2. What is the name of the thing that you are adjusting when you drag that central blue dot?

3. Use one of the outer blue dots to rotate the line. What happens to the equation as you rotate the line?

4. What is the name of the thing that you are adjusting when you rotate the line?

5. Use the drag and drop features of this app to produce lines with each of the following equations:

       a) $y = 2x + 4$ b) $y = -0.5x + 3$ c) $y = 3.25x - 6$

       d) $y = x - 7.33$ e) $y = 0.33x + 5.25$ f) $y = -4x - 2.75$

Center for Algebraic Thinking

Teaching Guide

Drag and Drop Linear Equations

App: Linear Model

Overview:

In this activity students use a tap-and-drag approach to linear equations as they manipulate a line and watch those manipulations change the equation of that line.

Common Core State Standards:
- CCSS.Math.Content.6.EE.C.9
- CCSS.Math.Content.8.F.A.1
- CCSS.Math.Content.8.F.A.3
- CCSS.Math.Content.8.F.B.4
- CCSS.Math.Content.HSA.CED.A.1
- CCSS.Math.Content.HSA.REI.D.10
- CCSS.Math.Content.HSF.IF.A.1
- CCSS.Math.Content.HSF.IF.C.7
- CCSS.Math.Content.HSF.LE.A.2

Encyclopedia of Algebraic Thinking:
- Analysis of Change: Understanding Slope
- Analysis of Change: Understanding the "b" in "y = mx + b"
- Patterns and Functions: Linear Function

Description:

The value of the app *Linear Model* lies in the ease with which students can manipulate a line simply by tapping and dragging blue dots that either translate or rotate the line. Because they can see corresponding changes in the equation for that line, as well as changes in a table of values, this application makes for a useful exploration tool. This activity allows students to become familiar with the ways in which they can manipulate a linear equation, and gives students a chance to connect those variations to specific vocabulary (y-intercept and slope).

Extensions:

- Multiple Representations of Lines

113

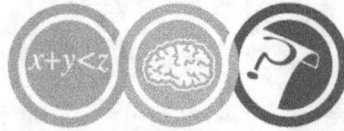

Center for Algebraic Thinking

Multiple Representations of Lines

For use with the iPad app *Linear Model*

Name: Date: Period:

In the "Function:" text input section, type in **y = 2x - 3**. Then tap 'Change'.

1. How many complete representations of this linear equation are there on the screen?

2. Describe each representation:

3. How can you use the table of x and y coordinate points to determine the slope of the line? Give an example using the equation above.

4. How can you use the equation to determine the slope and y-intercept of the line?

Center for Algebraic Thinking

Teaching Guide

Multiple Representations of Lines

App: Linear Model

Overview:

In this activity students use the iPad app *Linear Model* to explore three different ways of representing a line: as an equation, as a table of values, and as a graph.

Common Core State Standards:

• CCSS.Math.Content.6.EE.C.9 • CCSS.Math.Content.8.F.A.1 • CCSS.Math.Content.8.F.A.3 • CCSS.Math.Content.8.F.B.4 • CCSS.Math.Content.HSA.CED.A.1	• CCSS.Math.Content.HSA.REI.D.10 • CCSS.Math.Content.HSF.IF.A.1 • CCSS.Math.Content.HSF.IF.C.7 • CCSS.Math.Content.HSF.LE.A.2

Encyclopedia of Algebraic Thinking:
- Analysis of Change: Understanding Slope
- Analysis of Change: Understanding the "b" in "y = mx + b"
- Patterns and Functions: Linear Function

Description:

By exploring linear equations in a variety of ways, students will be better able to jump from one representation to another and gain useful knowledge from each. The ideal situation is for students to be able to start with one representation (either an equation, a table of values, or a graph) and produce the other two on their own. This activity gives students the opportunity to both focus on each representation individually and then make connections between the different representations. By making connections between the three representations, students will develop a fuller and more comprehensive understanding of linear equations.

Extensions:

Since this activity focuses on a single linear equation, teachers can extend the investigation by having students explore other linear equations that the teacher provides or that students themselves come up with.

Lion Grapher

Develop students' understanding of slope and y-intercept with this game environment which requires students to write equations to get a line into a lion's mouth. Multiple features in this great app that helps students use and construct meaning for all aspects of a linear equation. The app pushes students to understand how manipulation of the slope and y-intercept influence the graph of a line.

Center for Algebraic Thinking

Line-Building Challenge

For use with the iPad app *Lion Grapher*

Name: Date: Period:

*This is a **Screenshot Presentation** assignment. Here are your objectives:*

<u>Objective 1</u>

Using the Intercept Mode, create two of the following equations:

y = x	y = 2x + 3	y = -0.5x + 5
y = 1.3x - 7	y = -3x + 1.5	y = -2.25x + 6.5

<u>Objective 2</u>

Using the Point Mode, create two of the following equations:

y = x - 3	y = 7x - 7	y = 0.3x + 2
y = 5.75x - 9	y = -x + 1.5	y = 3x

<u>Objective 3</u>

Using the Line Mode, create two of the following equations:

y = 1.5x	y = 2.25x + 3	y = -0.5x - 8
y = 1.75x - 7	y = -3x + 9	y = 9x + 7

117

Teaching Guide

Line Building Challenge

App: Lion Grapher

Overview:
In this activity, students use the *Lion Grapher* app to construct lines in three different ways: by entering the y-intercept and a point on the line; by entering two points on the line; and by physically translating and adjusting the slope by touching and dragging.

Common Core State Standards:
- CCSS.Math.Content.6.EE.C.9
- CCSS.Math.Content.8.F.A.1
- CCSS.Math.Content.8.F.A.3
- CCSS.Math.Content.8.F.B.4
- CCSS.Math.Content.HSA.CED.A.1
- CCSS.Math.Content.HSA.REI.D.10
- CCSS.Math.Content.HSF.IF.A.1
- CCSS.Math.Content.HSF.IF.C.7
- CCSS.Math.Content.HSF.LE.A.2

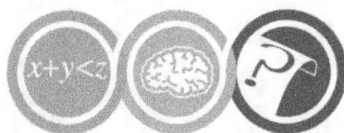

Encyclopedia of Algebraic Thinking:
- Analysis of Change: Understanding Slope
- Analysis of Change: Understanding the "b" in "y = mx + b"
- Patterns and Functions: Linear Function

Description:
By constructing lines in the three different ways described in the overview above, students will gain a deeper understanding of linear equations. In each mode, students are given six linear equations and must choose two of those to construct. Each mode allows students to focus on a different aspect of a line and its corresponding equation. By providing a variety of linear equations with differing levels of complexity, students with varying degrees of background knowledge will all be able to gain new understanding.

Extensions:
Since the activity asks students to construct two of the given six lines for each of the three objectives, teachers can extend this activity by having student construct all six of the lines for each objective. Students could also move on to the Lion Mode, in which students practice creating linear equations in a game context - they have to construct a line that passes through the head of lion that is placed somewhere on the coordinate system. Students have 200 seconds to complete this task as many times as possible.

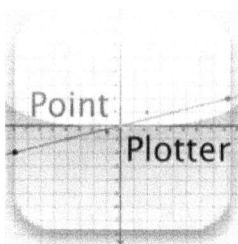

Point Plotter

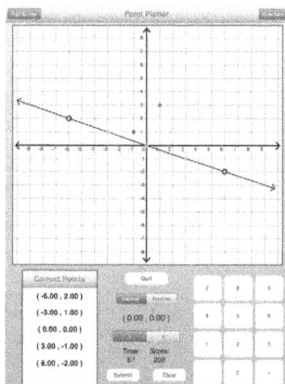

One misconception that students develop in algebra is that there are no points or a limited number of points between two points on a graph. This game uses students' knowledge of the 2D coordinate system and ability to recognize patterns to develop the concept there are an infinite number of points between two points. The premise is simple: you are shown a line segment, and you must identify as many points as possible that lie within that line segment. But you have just two minutes per round to do that. Submit enough correct points and you will make your way to the top of the High Scores let. With three difficulty levels you will be challenged no matter what your current skill level. The ultimate goal of this app is to give students the opportunity to discover for themselves the relationship that exists between all of the points that lie along a line. All correct points that a player submits are collected into a table. Players look at that table and try to describe how those points are related. Can you come up with a way to accurately   predict   other   points   that   lie   on   the   line?

Center for Algebraic Thinking

Moving from Points to Lines

For use with the iPad app *Point Plotter*

Name: Date: Period:

The challenge of the iPad app *Point Plotter* is to identify as many points as possible that fall within a particular line segment. In this assignment you will try to figure out how all of those points are related. Once you are skilled enough at *Point Plotter* to fill the "Correct Points" table with at least six points (at any difficulty level), start the investigation below.

Part 1: Graph and Table

1. After completing a *Point Plotter* game during which you identified at least six "Correct Points", graph the line from that game and record the correct points in the table below.

Correct Points (x , y)

Part 2: Searching for a Connection

1. Look at the table of correct points. Search for a connection between those points. Look for consistent changes as you move from one point to the next. In words, describe the connection that you see.

2. In this step you will describe that connection mathematically by trying to come up with an equation that will allow you to determine the y value of a point on the line if you are given the x value of that point. For each attempt below, write an equation for y, then test that equation on three correct points from the table on the previous page. To test your equation, start by copying the x value of a correct point into the x column of the table below, then calculate the y value using the equation you came up with, then add a check mark if the x and y values match the correct point from the previous page. Once you find an equation that works for all of the correct points, move on to the next problem.

Attempt 1: y = _____

x	y	Correct?

Attempt 2: y = _____

x	y	Correct?

121

Attempt 3: y = _____

x	y	Correct?

Attempt 4: y = _____

x	y	Correct?

Now that you have come up with an equation that relates the x and y values from the correct points table, you will use that equation to determine other points that lie along the same line. In the table below:

a) Chose an x value between -10 and 10 that is not already contained in the correct points table
b) Use your equation to calculate the y value that corresponds to that x value
c) Add that point to the graph on the previous page
d) Check to see if that point lies on the line
e) Repeat a) through d) three more times

x	y	On Line?

If all of the new points lie on the line, then congratulations, you have come up with an equation that accurately describes the line! If not, give it another shot with a new equation.

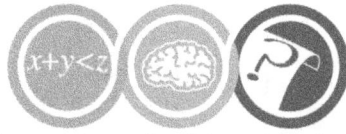

Center for Algebraic Thinking

Teaching Guide

Moving from Points to Lines

App: Point Plotter

Overview:

In this activity students develop the idea that a line is simply an infinite collection of related points, and are encouraged to develop the mathematical relationship between those points.

Common Core State Standards:

• CCSS.Math.Content.6.EE.C.9	• CCSS.Math.Content.HSA.REI.D.10
• CCSS.Math.Content.8.F.A.1	• CCSS.Math.Content.HSF.IF.A.1
• CCSS.Math.Content.8.F.A.3	• CCSS.Math.Content.HSF.IF.C.7
• CCSS.Math.Content.8.F.B.4	• CCSS.Math.Content.HSF.BF.A.1
• CCSS.Math.Content.HSA.CED.A.1	• CCSS.Math.Content.HSF.LE.A.2

Encyclopedia of Algebraic Thinking:
- Analysis of Change: Understanding Slope
- Analysis of Change: Understanding the "b" in "y = mx + b"
- Patterns and Functions: Linear Function

Description:

The *Point Plotter* app begins with nothing more than a line plotted on a coordinate system. The game aspect of the app involves students identifying as many coordinate points as possible that lie on that lie within a specific time limit. Once students can identify at least six correct points, they can begin this activity, which accomplishes two goals. The first goal is to show students that lines are nothing more than a collection of related points. As they get better at the game they will realize that there is no limit to the number of points that they could identify (if given an infinite amount of time). The second goal is for students to see the connection between those points. While this activity provides the space and probing questions to get at that connection, it leaves it for the student to develop and test those connections. That is the strength of this activity - it allows students to come up with their own ideas and then test those ideas by applying them to hard data.

Extensions:

- Moving from Points to Lines: The Follow Up

123

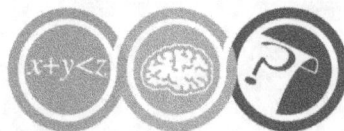

Moving from Points to Lines: The Follow Up

For use with the iPad app *Point Plotter*

Name: Date: Period:

In the first part of this assignment, you developed an equation that related all of the points that lie on a straight line. In this follow up assignment, we will take a closer look at equations that define straight lines - which are called **linear equations**.

Think back to the first part of this assignment... You may have noticed that as you move from one correct point to the next there is a consistent pattern to the changes in the x and y values of those points. In this activity we are going to develop that idea a bit further.

Part 1: Change in X & Change in Y
The first thing that we are going to do is look at how changes in y and changes in x are related as you move from one point on the line to another. To do that, start by picking two correct points from the table of correct values, and fill in the table below with the x value and y value of those points:

First Point	$x_1 =$	$y_1 =$
Second Point	$x_2 =$	$y_2 =$

Now we are going to look at how the y values change and how the x values change. Use the four values from above to fill in the table below:

Change in y	$(y_2 - y_1) =$
Change in x	$(x_2 - x_1) =$

In the final step, we are going to look at **the ratio** between the change in y and change in x:

Ratio of change in y and change in x	$(y_2 - y_1) / (x_2 - x_1) =$

That ratio that you found has a special name. It is called the **slope** of the line. Let's see what happens when we use two *different* points to calculate the slope. You will now complete the same steps for two different correct points. Select two new Correct Points and follow the same steps as before, and fill in the table that follows:

Third Point	$x_3 =$	$y_3 =$
Fourth Point	$x_4 =$	$y_4 =$
Change in y	$(y_4 - y_3) =$	
Change in x	$(x_4 - x_3) =$	
Ratio of change in y and change in x	$(y_4 - y_3) / (x_4 - x_3) =$	

If you did everything correctly, then the ratio you end up with when using the third and fourth points should be the same as for the first and second. Is that true for your points?

What this shows is that the **slope of the line** (which, again, is just the ratio you calculated) stays the same, no matter what two points you use to calculate it.

Part 2: The Y-Intercept
Most of the lines that we look at cross the y-axis in just one location (the only exceptions are vertical lines). The y value at that point where the line crosses the y-axis is called the **y-intercept**. What is the y-intercept for the line that you have been working with in this activity?

Part 3: The Equation
We will now bring the two concepts we have learned about (the **slope** and the **y-intercept**) together to create an equation for the line. The whole goal of writing an equation for the line is to make it easy for us to *calculate* the y value of the line after being given any x value. Since the y value of the line changes as we move left or right along the x-axis (unless the line is horizontal, in which case the line always has the same y value) then that means our equation should also show that same change.

To build our equation for the line, let's start by thinking about what is going on at the point where x is equal to zero. Since that is where the line crosses the y-axis, then the y value of the line is just equal to the y-intercept:

y = **y-intercept**   (when x equals zero)

Now if we think about moving left or right away from the point where x equals zero, then we can use our **slope** to figure out how much we need to add or subtract to the y-intercept. Recall that slope is just the ratio of the change in y value and the change in x value from one point to another on our line:

slope = (change in y) / (change in x)

We can solve that equation for (change in y) by multiplying both sides by (change in x):

slope * (change in x) = (change in y)

125

We can now add this change in y to our equation for y that we started on the last page:

$$y = \textbf{y-intercept} + \textbf{slope}*(\text{change in x})$$

But since we started the equation when x = 0, our change in x will simply be the x value at the new point:

$$y = \textbf{y-intercept} + \textbf{slope}*x$$

So once you know the y-intercept and the slope of the line, you can figure out the y value at any given x value using the equation above.

Our last step is to replace 'y-intercept' and 'slope' with letters so that it is easier to write. By convention we use the letter 'b' for 'y-intercept' and 'm' for 'slope':

$$y = b + m*x$$

And finally, by convention we switch the order of the terms on the right side of the equation:

$$y = \textbf{m}*\textbf{x} + \textbf{b}$$

That's it! We have finally reached the general equation of a line. The 'general' part means that any line has an equation that can be written in the form you see above. It took some work to build it up, but hopefully you now understand where it came from and see how useful it will be. To make sure it works for the line you've been working with in this activity, write the full equation for your line (by inserting values for 'm' and 'b'), and then use the equation to calculate a few points to see if they match up with your graph:

y = _____

x	y	On line?
-4		
-3		
-2		
-1		
0		
1		
2		
3		
4		

Teaching Guide

Moving from Points to Lines: The Follow Up

App: Point Plotter

Overview:

In this follow up activity to Moving from Points to Lines, students move from a specific line to the general equation for a line and are introduced to the terms 'slope' and 'y-intercept' as they build up the equation themselves.

Common Core State Standards:

• CCSS.Math.Content.6.EE.C.9	• CCSS.Math.Content.HSA.REI.D.10
• CCSS.Math.Content.8.F.A.1	• CCSS.Math.Content.HSF.IF.A.1
• CCSS.Math.Content.8.F.A.3	• CCSS.Math.Content.HSF.IF.C.7
• CCSS.Math.Content.8.F.B.4	• CCSS.Math.Content.HSF.BF.A.1
• CCSS.Math.Content.HSA.CED.A.1	• CCSS.Math.Content.HSF.LE.A.2

Encyclopedia of Algebraic Thinking:
- Analysis of Change: Understanding Slope
- Analysis of Change: Understanding the "b" in "y = mx + b"
- Patterns and Functions: Linear Function

Description:

By building on the concrete experience that students developed in the activity Moving from Points to Lines, this activity introduces linear equations in a constructive way that attempts to shed just as much light on the *process of doing mathematics* as the final ideas themselves (in this case, y = mx + b). The activity starts by explaining the concept of slope (using points that students recorded in the first activity), then moves to y-intercept, and finally to the general form of linear equations.

Extensions:

This activity contains quite a bit of information. A useful extension might be for students to condense all of this information while putting it into their own words, maybe in the form of a poster or some other presentation to give to classmates.

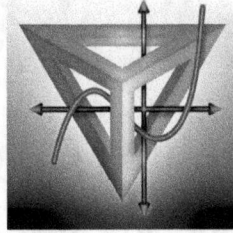

Math Flyer & Slope Slider

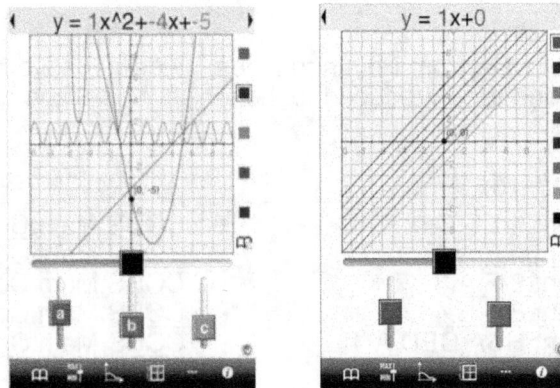

Traditional education uses static, motionless graphs to indicate the relationships between variables. While this works for some concepts, a student with a function and a picture of a graph gains no intuitive sense of the elements of the function and the relationship of each to the shape of the graph. With Math Flyer, a student can plot a graph and manipulate all of the variables and constants in that graph, allowing him or her to see the relationships firsthand. Math Flyer plots polynomials, all of the trig functions, exponentials, absolute value, square and saw waves, and so much more.

Center for Algebraic Thinking

Teaching Guide

Function Investigation

App: Math Flyer

Overview:

In the series of activities that follows this guide, students investigate functions by altering the coefficients in front of each term in the expression and then observing the effect that those alterations have on the graph of the function.

Common Core State Standards:
- CCSS.Math.Content.8.F.A.1
- CCSS.Math.Content.8.F.A.3
- CCSS.Math.Content.8.F.B.4
- CCSS.Math.Content.HSA.CED.A.1
- CCSS.Math.Content.HSA.CED.A.2
- CCSS.Math.Content.HSA.CED.A.3
- CCSS.Math.Content.HSA.REI.C.7
- CCSS.Math.Content.HSF.IF.A.1
- CCSS.Math.Content.HSF.IF.C.7
- CCSS.Math.Content.HSF.BF.B.3
- CCSS.Math.Content.HSF.LE.A.3
- CCSS.Math.Content.HSF.TF.B.5

Encyclopedia of Algebraic Thinking:
- Patterns and Functions: Transforming Functions
- Patterns and Functions: Graphing

Description:

In each of the activities in this series, students are instructed to adjust each coefficient (one at a time) of a particular function, and record their observations of how those adjustments affect the graph of the function. By manipulating the functions in this way, students will build up an understanding of the role of each coefficient (and the function as a whole) by direct observation. When they learn about functions in this way they will be more likely to remember the role of each coefficient. The functions that are included in this series are the following:

- Linear
- Quadratic (Part 1) [$y = ax^2 + bx + c$]
- Quadratic (Part 2) [$y = a(x - h)^2 + k$]
- Cosine
- Sine
- Absolute Value

Extensions:

Additional investigations are included on the last page of each of these activities. Teachers might also want to pair students together so that they can compare their observations and discuss any differences that they might find.

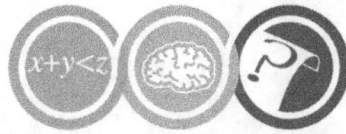

Function Investigation: Linear

For use with the iPad app *Math Flyer* or *Slope Slider*

Name: Date: Period:

Select the pre-made graph with the equation **y = mx + b**.

On the graph to the right, draw a sketch of the basic form of this function.

1. What happens when you increase and decrease the variable **m**?

2. Why do you think changes in **m** affect the graph in that way?

3. What happens when you increase and decrease the variable **b**?

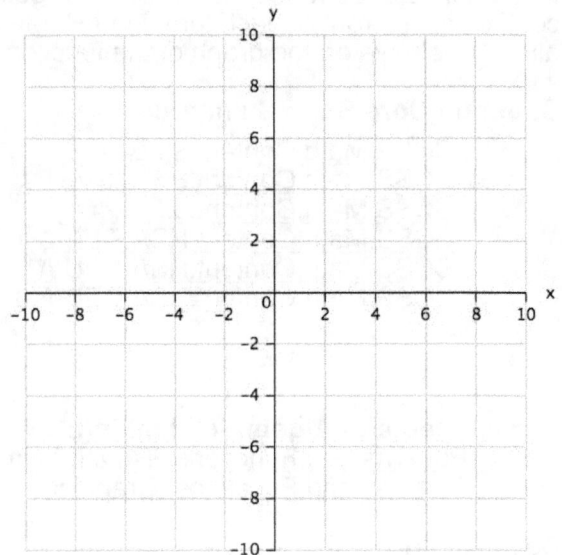

4. Why do you think changes in **b** affect the graph in that way?

Additional Investigations

1. What happens when you set **m** equal to zero?

2. What happens when you set **b** equal to zero?

Center for Algebraic Thinking

Function Investigation: Quadratic (Part 1)

For use with the iPad app *Math Flyer*

Name: Date: Period:

Select the Premade Graph with the equation $y = ax^2 + bx + c$.

On the graph to the right, draw a sketch of the basic form of this function.

1. What happens when you increase and decrease the variable **a**?

2. Why do you think changes in **a** affect the graph in that way?

3. What happens when you increase and decrease the variable **b**?

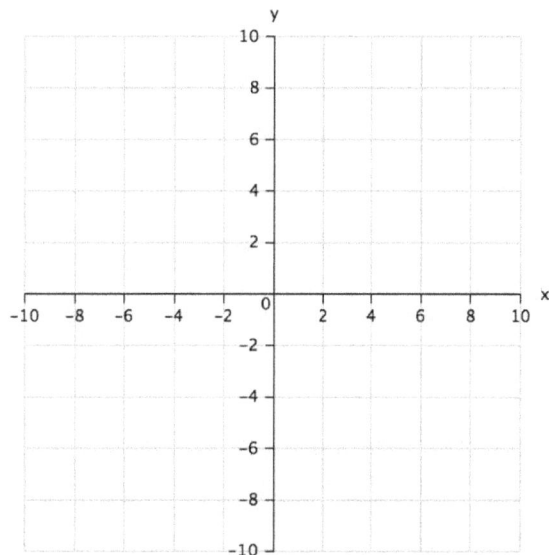

4. Why do you think changes in **b** affect the graph in that way?

5. What happens when you increase and decrease the variable **c**?

6. Why do you think changes in **c** affect the graph in that way?

131

Additional Investigations

1. What happens when you set **a** equal to zero?

2. What happens when you set **b** equal to zero?

3. What happens when you set **c** equal to zero?

Function Investigation: Quadratic (Part 2)

For use with the iPad app *Math Flyer*

Name: Date: Period:

Select the Premade Graph with the equation $y = a(x - h)^2 + k$.

On the graph to the right, draw a sketch of the
basic form of this function.

1. What happens when you increase and
decrease the variable **a**?

2. Why do you think changes in **a** affect the
graph in that way?

3. What happens when you increase and
decrease the variable **h**?

4. Why do you think changes in **h** affect the graph in that way?

5. What happens when you increase and decrease the variable **k**?

6. Why do you think changes in **k** affect the graph in that way?

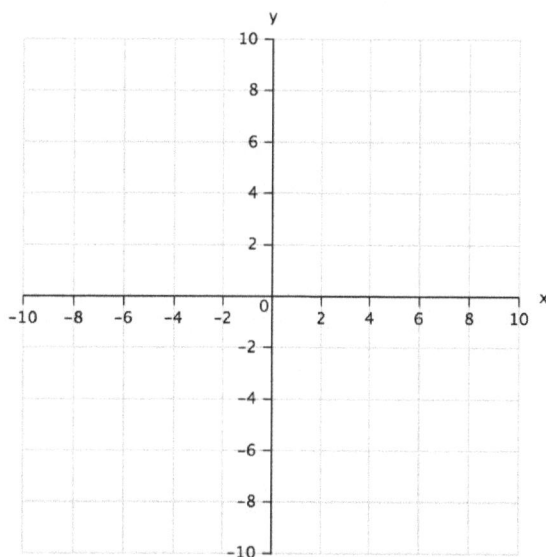

Additional Investigations

1. What happens when you set **a** equal to zero?

2. What happens when you set **h** equal to zero?

3. What happens when you set **k** equal to zero?

Center for Algebraic Thinking

Function Investigation: Cosine

For use with the iPad app *Math Flyer*

Name: Date: Period:

Select the Premade Graph with the equation **y = a*cos(bx + c) + d**.

On the graph to the right, draw a sketch of the basic form of this function.

1. What happens when you increase and decrease the variable **a**?

2. Why do you think changes in **a** affect the graph in that way?

3. What happens when you increase and decrease the variable **b**?

4. Why do you think changes in **b** affect the graph in that way?

5. What happens when you increase and decrease the variable **c**?

6. Why do you think changes in **c** affect the graph in that way?

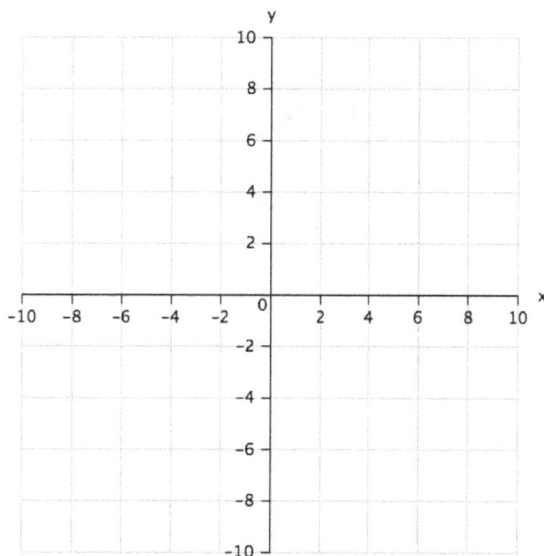

135

7. What happens when you increase and decrease the variable **d**?

8. Why do you think changes in **d** affect the graph in that way?

Additional Investigations

1. What happens when you set **a** equal to zero?

2. What happens when you set **b** equal to zero?

3. What happens when you set **c** equal to zero?

4. What happens when you set **d** equal to zero?

Function Investigation: Sine

For use with the iPad app *Math Flyer*

Name: Date: Period:

Select the Premade Graph with the equation $y = a*\sin(bx + c) + d$.

On the graph to the right, draw a sketch of the basic form of this function.

1. What happens when you increase and decrease the variable **a**?

2. Why do you think changes in **a** affect the graph in that way?

3. What happens when you increase and decrease the variable **b**?

4. Why do you think changes in **b** affect the graph in that way?

5. What happens when you increase and decrease the variable **c**?

6. Why do you think changes in **c** affect the graph in that way?

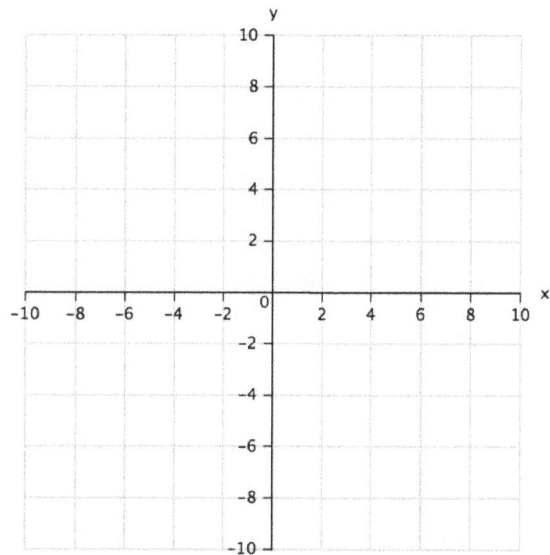

137

5. What happens when you increase and decrease the variable **d**?

6. Why do you think changes in **d** affect the graph in that way?

Additional Investigations

1. What happens when you set **a** equal to zero?

2. What happens when you set **b** equal to zero?

3. What happens when you set **c** equal to zero?

4. What happens when you set **d** equal to zero?

Function Investigation: Absolute Value

For use with the iPad app *Math Flyer*

Name: Date: Period:

Select the Premade Graph with the equation $y = a*abs(bx + c) + d$.

On the graph to the right, draw a sketch of the basic form of this function.

1. What happens when you increase and decrease the variable **a**?

2. Why do you think changes in **a** affect the graph in that way?

3. What happens when you increase and decrease the variable **b**?

4. Why do you think changes in **b** affect the graph in that way?

5. What happens when you increase and decrease the variable **c**?

6. Why do you think changes in **c** affect the graph in that way?

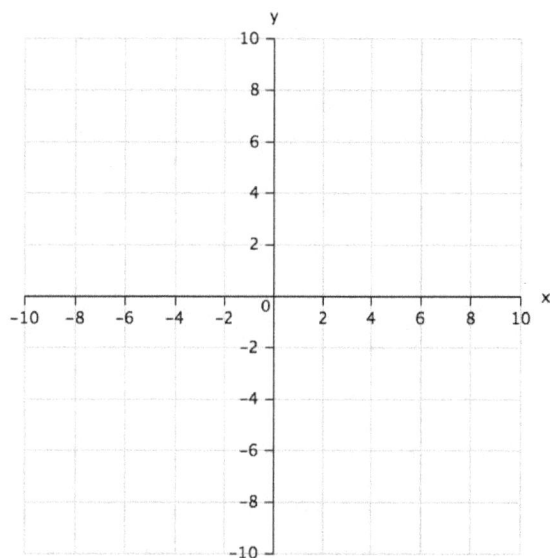

139

7. What happens when you increase and decrease the variable **d**?

8. Why do you think changes in **d** affect the graph in that way?

Additional Investigations

1. What happens when you set **a** equal to zero?

2. What happens when you set **b** equal to zero?

3. What happens when you set **c** equal to zero?

4. What happens when you set **d** equal to zero?

Graphing the Terms of Polynomials

For use with the iPad app *Math Flyer*

Name: Date: Period:

1. Use Math Flyer to graph the following four equations, all on the same screen:

a.) $y = 3$ b) $y = 2x$ c) $y = x^2$ d) $y = x^2 + 2x + 3$

2. Use the data table within the app or your own computing skills to complete the following table:

x	y = 3	y = 2x	y = x²	y = x² + 2x + 3
-3	3	-6	9	6
-2				
-1				
0				
1				
2				
3				

3. Now look at the graph of the four equations. What connection do you see between the graph of equation d) and the graph of the other three equations?

141

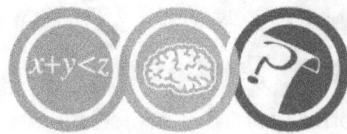

Center for Algebraic Thinking

Teaching Guide

Graphing the Terms of Polynomials

App: Math Flyer

Overview:

In this activity, students graph all terms of a polynomial individually, as well as the complete polynomial, in order to see how each term contributes to the function as a whole.

Common Core State Standards:
- CCSS.Math.Content.8.F.A.1
- CCSS.Math.Content.8.F.A.3
- CCSS.Math.Content.8.F.B.4
- CCSS.Math.Content.HSF.IF.A.1
- CCSS.Math.Content.HSF.IF.C.7
- CCSS.Math.Content.HSF.BF.B.3
- CCSS.Math.Content.HSF.LE.A.3

Encyclopedia of Algebraic Thinking:
- Patterns and Functions: Transforming Functions
- Patterns and Functions: Graphing

Description:
The *Math Flyer* app makes it easy to graph multiple functions on the same screen. In this activity, students use that ability to build up an understanding of how each term in a polynomial contributes to the graph of that polynomial. After graphing each of the terms of a polynomial, as well as the complete polynomials, students then complete a table of values for each of the terms to reinforce the idea that the picture of the complete polynomial is simply the addition of all of the terms of that polynomial. By seeing this in two different representations (graph and table), students will be more likely to identify this relationship on their own and more likely to remember it in the future.

Extensions
Teachers can extend this activity by giving students additional polynomials and asking them to analyze the new polynomials in the same way that they analyzed the polynomial in this activity.

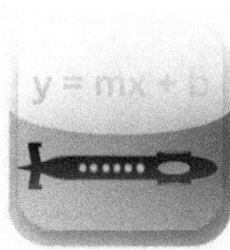

Submariner Algebra

Develop students' understanding of points on a line, slope, and y-intercept with this game environment. Similar to the game "Battleship", students graph submarine paths (3 lines) for a competing student to find. Dropping depth charges (x,y points), students seek the submarine paths. After a couple of hits, students use the points to identify the equation of a line to determine if they have found one of the submarine paths. The app pushes students to understand how manipulation of the slope and y-intercept influence the graph of a line.

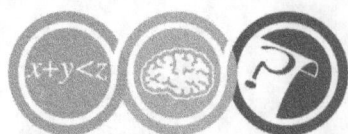

Submariner Algebra Workspace

For use with the iPad app *Submariner Algebra*

Name: Date: Period:

Use the space below to show the work needed to calculate the slope and y-intercept of the enemy's submarine route after you have gotten two hits.

Round One:

Hit #1: (_____ , _____)   Hit #2: (_____ , _____)

Workspace:

Slope: Y-Intercept:

Round Two:

Hit #1: (_____ , _____)   Hit #2: (_____ , _____)

Workspace:

Slope: Y-Intercept:

Round Three:

Hit #1: (_____ , _____)   Hit #2: (_____ , _____)

Workspace:

Slope: Y-Intercept:

Round Four:

Hit #1: (_____ , _____)   Hit #2: (_____ , _____)

Workspace:

Slope: Y-Intercept:

Round Five:

Hit #1: (_____ , _____)   Hit #2: (_____ , _____)

Workspace:

Slope: Y-Intercept:

145

Teaching Guide

Submariner Algebra Workspace

App: Submariner Algebra

Overview:

This activity provides a space for students to show their work as they play the single-player version of the game *Submariner Algebra*, in which they must determine the line that an enemy submarine follows after locating two points that the submarine crosses (with a gameplay that resembles the board game Battleship).

Common Core State Standards:

• CCSS.Math.Content.7.EE.B.4	• CCSS.Math.Content.HSF.IF.A.1
• CCSS.Math.Content.8.EE.B.5	• CCSS.Math.Content.HSF.IF.B.4
• CCSS.Math.Content.8.F.A.1	• CCSS.Math.Content.HSF.BF.A.1
• CCSS.Math.Content.8.F.B.4	• CCSS.Math.Content.HSF.LE.A.2
• CCSS.Math.Content.HSA.CED.A.1	• CCSS.Math.Content.HSF.LE.B.5
• CCSS.Math.Content.HSA.REI.B.3	

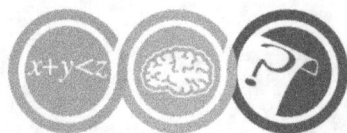

Encyclopedia of Algebraic Thinking:
- Analysis of Change: Understanding Slope
- Analysis of Change: Understanding the "b" in "y = mx + b"
- Patterns and Functions: Linear Function

Description:
The app *Submariner Algebra* does not require a paper resource to accompany it, but this resource serves two functions. The first is to encourage students to work through the calculation of slope and y-intercept by hand, rather than trying to do it in their head based on the points that they have identified as hits. This reinforces the good habit of showing your work. The second function is to provide accountability - the teacher can collect this record of students' work at the end of the activity to see both the quality of the students' work and the quantity - how many rounds students were able to complete during the allotted time for this activity.

Extensions:
If you have multiple iPads in the classroom, students can use this resource in the 2-Player Mode to challenge each other. You could even set up a tournament so that each student plays multiple rounds against various opponents.

Center for Algebraic Thinking

Submariner Algebra: The Paper Version

For use with the iPad app *Submariner Algebra*

Name: Date: Period:

In the iPad app Submariner Algebra *you battle your opponent to see who can determine the path of the opponent's submarine first. In this activity you will bring the battle into the real world.*

Instructions

Setting Up

(1) Make sure that your opponent cannot see your playing board (on the next page) and that you cannot see theirs.

(2) Select the path of your submarine, and write the equation for that path on the following page.

(3) Draw the path of your submarine on the coordinate system.

(4) Decide what marks you want to use for your opponent's guesses, your incorrect guesses, and your correct guesses. Add each mark to the **KEY** for reference.

Gameplay

(1) You and your opponent alternate guesses, adding the appropriate marks to your coordinate system for each guess - a hit or a miss.

(2) Once you have two hits, determine the path of your opponent's submarine, and state the equation of the path to your opponent to see if you are correct or not.

(3) First team to correctly guess the path of their opponent wins!

147

Your Submarine's Path: y = _____

KEY Opponents Guesses: ____   Your Incorrect Guesses: ____   Your Correct Guesses: ____

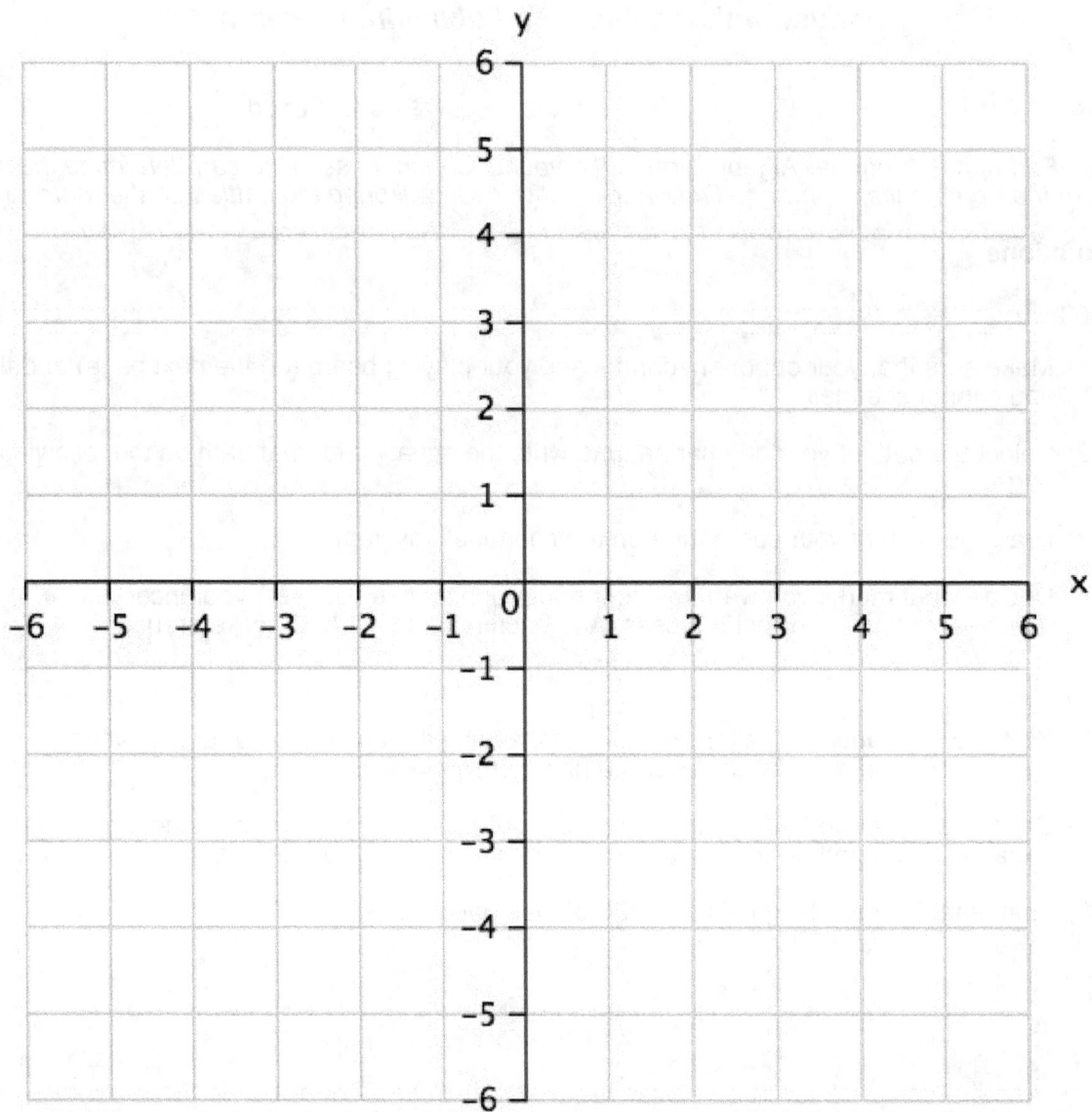

Workspace:

Your Opponent's Path: y = _____

Center for Algebraic Thinking

Teaching Guide

Submariner Algebra - The Paper Version

App: Submariner Algebra

Overview:

In this activity students follow the same gameplay as that found in the iPad app *Submariner Algebra*, but they play the game on paper instead of an iPad.

Common Core State Standards:

• CCSS.Math.Content.7.EE.B.4	• CCSS.Math.Content.HSF.IF.A.1
• CCSS.Math.Content.8.EE.B.5	• CCSS.Math.Content.HSF.IF.B.4
• CCSS.Math.Content.8.F.A.1	• CCSS.Math.Content.HSF.BF.A.1
• CCSS.Math.Content.8.F.B.4	• CCSS.Math.Content.HSF.LE.A.2
• CCSS.Math.Content.HSA.CED.A.1	• CCSS.Math.Content.HSF.LE.B.5
• CCSS.Math.Content.HSA.REI.B.3	

Encyclopedia of Algebraic Thinking:
- Analysis of Change: Understanding Slope
- Analysis of Change: Understanding the "b" in "y = mx + b"
- Patterns and Functions: Linear Function

Description:
There are a variety of reasons that you might want use the paper version of this game. One is that you might not have enough iPads for all students. With this paper version you could have the entire class playing at once, some with iPads, others with pencil and paper. Another reason is because of the added responsibilities on the student when playing on paper. Rather than just tapping a location, students will have to list a coordinate point, and respond to the guesses of their opponent. They will also have to evaluate the correctness of the path that their opponent guesses, maybe recognizing that they determined the y-intercept correctly but didn't get the slope right. Even the simple act of letting students chose their own marks for hits and misses could lead to increased student engagement during the activity. For these reasons it might be worth giving this activity a try in conjunction with the iPad version.

Extensions:
After a few rounds of play you could have students complete the Developing Game Strategies activity, and then discuss their approaches with each other.

149

Tortoise & the Hare Algebra

One of the most challenging topics to understand in algebra is rate of change. This app helps students explore the effects of different rates of change on the classic race between the tortoise and the hare. Students can manipulate how many feet per second each racer travels, including a little nap time for the hare and then watch the race be animated. Students can also change perspectives between seeing the overall map of the race and watching the individual racers up close and see how the rates of change for each racer look in each context.

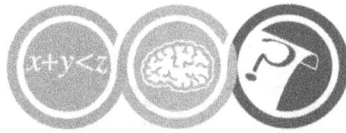

Center for Algebraic Thinking

Tortoise & the Hare Algebra Challenge
Progress Monitor

For use with the iPad app *Tortoise and the Hare Algebra*

Name: Date: Period:

Use this sheet to keep track of your progress as you complete the in-app challenges.

	Solution	Check
Level: Challenge:		
Level: Challenge:		
Level: Challenge:		

151

	Solution	Check
Level: Challenge:		
Level: Challenge:		
Level: Challenge:		
Level: Challenge:		
Level: Challenge:		

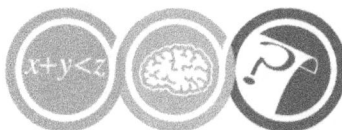

Teaching Guide

Tortoise and the Hare Algebra
Progress Monitor

App: Tortoise and the Hare Algebra

Overview:

This resource is a way for students to keep track of their progress in the 'Challenge' section of the iPad app *Tortoise and the Hare Algebra*.

Common Core State Standards:

- CCSS.Math.Content.8.F.B.4
- CCSS.Math.Content.HSA.CED.A.1
- CCSS.Math.Content.HSA.REI.B.3
- CCSS.Math.Content.HSA.REI.C.6
- CCSS.Math.Content.HSF.BF.A.1
- CCSS.Math.Content.HSF.LE.A.2

Encyclopedia of Algebraic Thinking:

- Analysis of Change: Understanding Speed as Rate of Change
- Modeling: Translating Word Problems into Equations

Description:

The challenges provided in this app ask a variety of questions about particular scenarios that would play out as the tortoise and the hare race at different speeds. The beauty of the app is that no key is needed - students can simply input the conditions, tap 'Start Race,' and then watch to see if the results confirm that they answered the challenge correctly. This paper resource allows students to keep track of which challenges they have completed successfully as well as record their solutions and the process of arriving at that solution.

Extensions:

Teachers could extend this activity by having students *create their own challenges* that either they themselves or a classmate has to solve.

www.ingramcontent.com/pod-product-compliance
Lightning Source LLC
LaVergne TN
LVHW081345060426
835508LV00017B/1430

9780692468241